ANIMATE EARTH

Science, Intuition and Gaia

D1382683

Stephan Harding

Foreword by Lynn Margulis

For Lisa and Roger
My new-found friends
in Gaia.
with love
Stephen

 A Sciencewriters Book

CHELSEA GREEN PUBLISHING COMPANY
WHITE RIVER JUNCTION, VERMONT

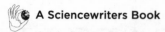 **A Sciencewriters Book**

scientific knowledge through enchantment
Sciencewriters Books is an imprint of Chelsea Green Publishing. Founded and codirected by Lynn Margulis and Dorion Sagan, Sciencewriters is an educational partnership devoted to advancing science through enchantment in the form of the finest possible books, videos, and other media.

Illustrations and cover artwork by Deirdre Hyde, www.deirdre-hyde.com
Diagrams by Rick Lawrence

Printed in the United States
First printing, September 2006
10 9 8 7 6 5 4 3 2 1

First published in the United Kingdom in 2006 by Green Books Ltd.

Our Commitment to Green Publishing
Chelsea Green sees publishing as a tool for cultural change and ecological steward-ship. We strive to align our book manufacturing practices with our editorial mission, and to reduce the impact of our business enterprise on the environment. We print our books and catalogs on chlorine-free recycled paper, using soy-based inks, whenever possible. This book may cost slightly more because we use recycled paper, and we hope you'll agree that it's worth it. Chelsea Green is a member of the Green Press Initiative (www.greenpressinitiative.org), a nonprofit coalition of publishers, manufacturers, and authors working to protect the world's endangered forests and conserve natural resources.

Animate Earth was printed on 55# Cascade Enviro ED100, a 100 percent post–consumer waste recycled, old growth forest–free paper supplied by RR Donnelley.

Library of Congress Cataloging-in-Publication Data is available from the publisher.

Chelsea Green Publishing Company
Post Office Box 428
White River Junction, Vermont 05001
(802) 295-6300
www.chelseagreen.com

Contents

DEDICATION

For Edda and Severin, and Oscar and Julia
and
For the wild, sacred beauty of the Sengwa River
and Ntaba Mangwe, Mountain of Vultures

Acknowledgements

To the people who nurtured me and brought me up with the freedom to explore my deep love of nature I owe the deepest gratitude: my father Severin, my mother's sister Hanka, my foster-mother Julia Rodriguez, my stepmother/sister Lucy and her mother Meg. And now, as a grown man, my gratitude and love go out to Julia and little Oscar, who, turned just four, suggested that I call this book *Being Gaia*, or *Desperate Earth*—now one of the chapter titles.

Of all the astonishing people who have helped me grow intellectually and spiritually I have space to thank but a few—Andrew Proudfoot, John Eisenberg, Deirdre Hyde, Bob Carlson, James Lovelock, Lynn Margulis, Arne Naess, Satish Kumar, Helena Norberg-Hodge, John Page, Peter Bunyard, Fritjof Capra, Brian Goodwin, Henri Bortoft, Margaret Colquhoun, Patsy Hallen, Peter Summers, Jonathan Horwitz, Per Espen Stoknes and Per Ingvar Haukeland. To David Abram I owe a great debt of gratitude for his inspiring teaching and writing. It is from him that I imbibed the concept of 'animate earth' that eventually surfaced as the title of this book. All these friends and many others have added so much to the kaleidoscope of understanding that I attempt to give back to the world through the service of speaking and teaching.

Next, my sincere thanks to everyone who read the manuscript, entirely or in parts. I am deeply grateful to James Lovelock, my friend and long-standing scientific mentor, for a careful and critical reading of the entire text that has kept the science straight and convinced me that writing this book was worthwhile. My friend the brilliant Norwegian ecopsychologist Per Espen Stoknes made sure that the 'soul' dimension of the narrative was not lost to sight. Peter Reason from the University of Bath also read the entire book, and urged me not to lose sight of the animistic perspective that he found so distinctive about this work. Galeo Saintz, a recent Schumacher College MSc graduate, kindly read the first draft and made valuable comments, as did Mel Risebrow and Harriet Posner. Ed O'Sullivan read the whole book and was greatly encouraging. Henri Bortoft checked the philosophical sections and Brian Goodwin, my friend and mentor, read enough to give me the confidence to continue. John Page, my good friend of many years who has an eye for good writing, surprised me by telling me that he enjoyed what he read. David Abram made very helpful suggestions on the first two chapters. I am very grateful to Liz Morrison for her valuable comments on the experiential aspects of the text.

I also thank Schumacher College, my second *alma mater,* and Anne Phillips, its Director, for allowing me to develop my understanding of Gaia in my own way without any pressure to conform to any standards, without formal assessment, and with the only expectation that I would teach and take people out to the woods and to the sea. To have been given such freedom is a rare honour in these troubled times in the world of education. I would like to acknowledge the support and inspiration of Brian Goodwin, Terry Irwin and Gideon Kossoff, my colleagues on the MSc in Holistic Science.

My gratitude to Richard and Diana St Ruth for giving me a lovely room in the Golden Buddha Centre on the edge of Dartmoor during the winter of 2005 where I wrote for five quiet weeks. Without the calm of their wonderful retreat centre and their generous friendship this book could not have been completed. I am also deeply grateful to Derek Hook, Gary Primrose and Christina West at Yewfield in the English lakes for enduring my family for two weeks whilst I mused on various aspects of the work, and to Andy and Sarah Middleton and their charming children for their generous hospitality at Whitesands Bay, and for urging me to write in the splendid tower of their lovely hotel in St. David's.

My special thanks to the artist Deirdre Hyde, my long-standing friend whom I met whilst I lived in Costa Rica, for so generously painting the cover picture and the wonderful illustrations that grace this book.

Lastly, my thanks to John Elford of Green Books for his brilliant editorial skills and for his fine handling of this book.

Foreword
by Lynn Margulis

"Dghem," however pronounced, Lewis Thomas (physician-writer, 1913-1993) reminds us, is the ancient Indo-European term for "Mother Earth." Surprisingly different English words derive from *dghem*. This fact illuminates Stephan Harding's admonition that we humans can lead happier, healthier, more knowledgeable and lively lives if we return to the "experience of Gaia as a living presence" (p.41). In *Animate Earth* our author, an Oxford University PhD zoologist who has taught for 15 years at Schumacher College in the southwest corner of England, near Totnes, Devon, begins with prehistory. He writes that

> traditional peoples all over the world have believed in an Earth mother who bestows life and receives the dead into her rich soil. The ancient Greeks called her Gaia, the earthly presence of *anima mundi*, the vast and mysterious primordial intelligence that steadily gives birth to all that exists. The great nourishing subjectivity—at once both spiritual and material—that sustains all that is.

Such ancient knowledge reveals itself in the *dghem* word that somehow led to the word "Earth" in English and related words like *Erde* in Germanic languages. Harding revels in a comparable term, *"anima mundi,"* or in translation from Latin, *"the soul of the Earth."*

The living, animate Earth idea continues to influence us today through a circuitous Platonic route: in his *Timaeus* Plato claims "The world is indeed a living being supplied with soul and intelligence... a single visible entity, containing all other entities" (p.24). And from *dghem,* the Earth-word, comes, for example, "humus," "humic acid," "humor," "humanity," "humanist," "humanities," "humility," and most telling, "human." Harding, in luminous prose and with scientific passion shows us that, from the beginning of written documents, and deep in our intuitive consciousness, we members of the genus touted as "wise men," *Homo sapiens*, with our literally fantastic capacity for denial and self-deception, know, really know, that we belong to the Earth. No matter our religious preconceptions or our particular straightjacket of socialization, of so-called education, we recognize emotionally and intellectually that not only do we come from the prodigious Earth, but that, alas, when our heart stops beating each of us will return to her. Most of us, indeed, those of us who do not perish over the ocean in a misguided airplane, fall victim in war to the ravages of bombs, or succeed in the mandate of personal cremation as an ecologically-friendly alternative, we will, as the ancients proclaimed, be received as her dead into Mother Earth's "rich soil."

Harding, who is profoundly educated in the biological sciences, especially zoology, has left the halls of academe far behind. He has transcended his own superb Oxonian background. His intellectual excursion has rejected that English-speaking, Anglo-American, Darwin-distorting, population-genetics-capitalistic, idiosyncratic "evolutionary biology," I see mainly as a subfield of loyalists. The blinkered abomination of those who call themselves "evolutionary biologists" permeates the great universities of California, eastern Canada, England, New England, the mid-Atlantic states, and many other former British colonies. They control the "organismic and evolutionary biology" (OEB) budgets. Harding doesn't practice such "academic apartheid" (Lovelock's term). No, his loyalty is to the whole living Earth, and yet he never waivers in his dedication to science as a way of knowing. This contribution, *Animate Earth*, because of the accessibility of its language and in spite of the depth of its erudition, is as laudable as it is unique.

"Science is the search for truth," said David Bohm, quantum mechanical theoretical physicist, "whether we like it or not." Bohm (1917-92), who worked for years in England and Israel, was born in Pennsylvania and studied in this country with Albert Einstein and J. Robert Oppenheimer. In his 30's, as a victim of McCarthyism, he abandoned his US residency for the rest of his life; he went first to Brazil and, in the end, died in London. For Harding, Bohm's "truth" is as illusive as the biologist's "life itself," yet "science" really *is* the search for, not necessarily the discovery of, truth.

Harding has no difficulty with "primary science," the immense literature of "scientific truths" that mightily attempt to explain nature in, as close as possible, her own language. Our author greatly respects the mode of learning about the world called primary science where the discoveries are recorded by the scientists themselves. He honors practicing scientists as he informs himself of their hard-earned arcane knowledge. However, he transcends the practitioners. Because the entire Earth and its history is Harding's object of study he never claims that any authentic scientific observation or measurement is "out-of-his-field." Whether a meteorologist who tracks climate change, a hammer-toting geologist who reconstructs (from the chemical state of minerals trapped in sedimentary rocks) the transition from an oxygen-poor to an oxygen-replete atmosphere, or a behavioral biologist who describes choreographically the mother-toddler bond, Stephan reads their work with a scientist's critical eye.

Scientifically eclectic, Harding's view is inclusive, tolerant, and enlightened as are those of many "deep ecologists" and "environmental activists." However, no other zoologist is like him in my experience. Superbly, if peculiarly, educated in the best British tradition but tutored by James Lovelock himself in rigorous, iconoclastic scientific analysis, Harding's insights are tempered by personal experience in the wilds of South America and in Africa. I suspect that no one else in the English-speaking world has the personal field knowledge, rigorous Lovelockian quantitative scientific training, intellectual curiosity, and years of response to earnest questions by curious, talented, international, and often very mature students, especially those attracted to Schumacher College. No "straight-A" academic-type could even conceive of this book, let alone write it!

I am reminded of the deep thinking of James Baldwin, the African-American writer who, himself disillusioned by the entrenched delusions and intolerant power structure, could not continue to reside in his own beloved country. Baldwin, in a recorded interview in Amherst, Massachusetts, where I am now, said:

> It is hard to think of America as a culture. It impresses me as being a conglomeration of many cultures, none of them really respected. All of them at the mercy of what this country imagines itself to be. What this country imagines itself to be may be exactly what it is: what it imagines itself to be, one would have to conclude from what it says, is a collection of pragmatic, pious businessmen. That seems to be the American self-image. Nothing could be more sterile. And for any artist finding himself or herself in the middle of a terrifying sterility there is very little that is admired in this country that anyone can use. It seems to me, you know, the chimera of success. You have to be nurtured that way, so I can't answer what it says about the culture except that—because one doesn't want to say that

its sterile—the culture. Its sort of impossible to imagine a sterile culture. It is a contradiction in terms. It seems to me there's something buried here, buried alive trying to, trying…, struggling for expression. I may be wrong about this but it seems to me that key to American life—if one can say a key to anything so vast as that generality—the key to American life seems to me to be involved with their stubborn, manic refusal to accept their history. The history that is taught, the history that is promulgated in all the schools, in all our institutions is not true. Its simply not true that the country was founded by freedom-loving heroes. The myth of George Washington is not true. None of it is true. The Declaration of Independence was signed by slave owners.

It is the "stubborn, manic refusal to accept" not United States of America written history but worldwide *natural history* that has promulgated the need, on a global scale, for the antidote of Harding's "*Animate Earth.*" Science textbooks, especially those on life sciences, general biology, or evolutionary biology do not teach science. They teach common myths of nation, profession, and "Western civilization," myths that conform to current budgetary needs of politically powerful academic departments, international corporations in need of technicians, greed-driven book publishers, and others. The teachers, even more than the students, are victims. Recent science textbooks in no way bring us closer to the hard-earned scientific truths like those expounded here by Stephan Harding. I suggest that the textbooks provide us with a wonderful Baldwinian illustration of our "stubborn, manic refusal to face" not only our national but our natural history.

The history of life as evolutionary narrative, told in Harding's book with sensibility, is entirely absent in life sciences textbooks that feel compelled to "cover the material." Such books that drone on about animal physiology, cell and molecular biochemistry, anatomy and the like are often opaque and confusing not only to students but even to the teaching faculty. General biology books tend not only to be ignorant of the most basic geological facts but are deficient in understanding the microbiological world that supports all life on Earth. By contrast, the microbiology texts are in frank contradiction to those of "general biology": botany (the study of plants), mycology (the study of fungi such as molds, mushrooms and yeasts), and protoctists. Often the evolution books ignore sciences critical to their task, such as atmospheric chemistry, chemical thermodynamics, and microbial ecology. Rather, in the practice of population ecology and population genetics they glorify a bizarre numerology, a cryptic mathematics that proffers unmeasurables such as "reciprocal altruism" and "inclusive fitness." These unquantifiables are inappropriate to our cell-biological and other chemically-based subject matter, in my view.

Harding's unique, hypocrisy-free narrative probably can never be sold in bulk. It will probably not be embraced by the academic establishment in any Anglophone country but, like other small, clearly written books with literary merit, *Animate Earth* is a text to be studied by curious bibliophiles, students and teachers especially of biogeological topics. It *should* serve as required reading for at least the following kinds of classes (in rough order of importance to their required subject matter: evolution, Earth sciences, biology, sedimentology, stratigraphy, palaeontology, economics, comparative religion, ethics, sociology, physiology, meteorology, and especially philosophy.

The Baldwinian "manic refusal to face our history"—and our natural history—is dwarfed by our "manic refusal" to even admit the existence of philosophy. Most citizens of the United States don't see any value in any philosophy. They don't think they have, or need, any philosophy. And since Harding's book is one of philosophy consistent with scientific observation, its subject matter eludes easy market classification.

Harding's great merit, it seems to me, is his provision of a factual foundation to Whiteheadian philosophy. *Animate Earth* should be translated into all the major languages of the world. Why? Because of its fidelity to the best available truths of science, rather than to any academic, corporate or political constituency. Unlike the usual science books, in *Animate Earth* Harding incorporates into his work the cunning of that particular branch of animal behavior that we name human psychology, with its insights into emotional inference, feigning, betrayal, intuition, delusion, persuasion, tribal loyalty and the like. Harding faces immense odds against him in this realm of "manic refusal," this reductionist, mercantile world. I salute his valiant, and mostly successful attempt to "keep things whole."

At the end of this delightful narrative about the trials and tribulations on the Earth's surface, you, the reader, will understand Gaia. You will know for yourself that what has been called "the Earth's environment" is no externality. The environment is part of the body. Therefore, for us, the talkative, lying, quarrelsome but endlessly manipulative, social ape, the disrespectful act of despoilment, the self-mutilation, the pandemic we call progress (e.g., deforestation, desertification) are, for Gaia, only petty activities, a masochism writ large of the mammalian kind that Gaia has seen before. Gaia continues to smile: *Homo sapiens*, she shrugs, soon will either change its wayward ways or, like other plague species, will terminate with a whimper in the current scourge, in this same accelerated Holocene extinction it initiated and has sustained over the past 10,000 years.

LYNN MARGULIS
Distinguished University Professor
in the Geosciences Department
University of Massachusetts, Amherst
June 2006

Introduction

Nature and the countless ways we have discovered to reproduce it, visually, aurally and verbally, are antipathetic; above all I believe it is so verbally. Words cannot reproduce nature; they exist in totally different worlds. John Fowles

I embarked on the writing of this book thanks to the persistent encouragement of Satish Kumar, my colleague at Schumacher College, and John Elford at Green Books, both of whom convinced me that it was important to communicate Gaian ideas and intuitions by as many means as possible at this time of severe ecological crisis and social meltdown.

The translation from the oral to the written mode of communication has been a challenging task. I find it easier to talk about our animate Earth, much as a bushman storyteller spins tales about the magical beings that made the world to a small group of fellow humans under the stars in some remote area of wild country. Gaia awakens for me when I dance her complex cycles and feedback loops into life, when I write her chemistry on a board with adoring hands, and when I see her sparkling in the eyes of an eager audience. Bringing her to life through the medium of the written word has been far more demanding.

This is at least in part a book about science, which is concerned with *explanation*. Scientists deal with facts, models and predictions, and enjoy the feeling that a hypothesis has been validated by means of empirical research. Explanation is the essential and vitally important work of the rational mind, but we must not lose sight of an equally important need for *understanding*, for contact with the realm of meaning, where we seek intimacy and connection with what has been explained.* In the way I use it

---

* This distinction between *understanding* and *explanation* comes from the German philosopher Wilhelm Dilthey (1833–1911).

here, understanding is not concerned with telling us how a thing has come to be or how it works—it seeks only empathy and a sense of mystery. Explanation is rational; understanding is intuitive. Reconnecting these two severed branches of our psyche is a vitally important task if we are to respond appropriately to the vast ecological crisis that our culture has unleashed upon the world. This book is no more than a simple attempt to make some headway in this direction.

Wherever possible, I have tried to bring explanation and understanding together by placing the concepts of Gaian science as firmly as possible within the realms of meaning and directly felt experience. In part, I have done this by telling stories about the scientific 'invisibles'—the atoms, microbes and feedback relationships that make up the astonishing body of our living Earth—but in the end, the deeper meaning contained in this book cannot do its work unless you digest what you have read and make it your own, under the stars perhaps, or by the roaring sea.

About the text in italics

The sections in italics (between grey bars) contain text that embodies an intuitive, experiential approach to Gaia. Some are contemplative, others give you things to think about, others relate my own personal deep experiences of the natural world, and yet others encourage you to be reflective about Gaia.

You can work with the contemplative approaches in various ways. The most basic is simply to read them almost as a kind of poetry. Better than this is to read each paragraph, and then close your eyes and spend time dwelling on the images so that your intuition and feeling are given the chance to work on them. Lastly, and perhaps best of all, is to make a recording of yourself reading out one of these passages, and to listen to the recording out in nature, as a kind of guided visualisation. Make sure to leave plenty of silence between the paragraphs, and feel free to change and embellish the material as you will; and better still, to experiment with your own approaches.

Chapter 1

Anima Mundi

Sea Change.

Reason flows from the blending of rational thought and feeling. If the two functions are torn apart, thinking deteriorates into schizoid intellectual activity and feeling deteriorates into neurotic life-damaging passions. Eric Fromm

Muntjac

Our world is in crisis, and, regrettably, our way of doing science in the West has inadvertently contributed to the many problems we face. I began to realise that something was seriously amiss with our mode of scientific enquiry when I was 25 years old. I had just come back to England after three years away as an ecologist and teacher in Venezuela and Colombia. Feeling my usual urgent need to connect with nature, I had lost no time in finding a quiet wood, which to my delight, was peppered with the tracks of tiny cloven-hoofed beings. But whose tracks were these? Fascinated, I had hidden myself in a thicket overlooking a broad woodland path, waiting for the mysterious creatures to appear. As the sun settled on the horizon and dusk bathed the wood in a deep purple light, a tiny deer stepped out of the trees and stood out in the open, a creature so small that it was more like the duikers I had occasionally spied in the wild bush country of Zimbabwe than any deer native to the British Isles. The little creature exuded a deep peace and an easy elegance that totally captivated me, transforming the whole wood. In the presence of this being, a profound sense of the inexpressible beauty of nature wafted over me like subtle smoke, enveloping me in a feeling of deep peace and happiness.

The little deer was a Reeve's muntjac deer (*Muntiacus reevesi*), a relatively recent addition from southern China to the fauna of the British Isles, and one of the world's smallest deer, with a shoulder height of only 45-50 centimetres. I cycled back to the wood many times to see muntjac. It was during one of these visits that I was gripped by the idea of devoting the next few years of my life to delving into the lives of these enigmatic creatures. Soon after, to my delight, I was given the chance to do my doctorate on muntjac ecology and behaviour at one of the world's very best zoology departments, at the University of Oxford.

It was hard to find a good study area. For a whole year I laboured in vain in a dense thicket behind an army barracks trying to observe muntjac behaviour, but the best I could do was to take plaster casts of hundreds of muntjac footprints. Scientifically, these were almost worthless, but collecting them had at least kept me busy. Then of course, there was the inevitable collection of muntjac dung, which would at least yield some interesting information, despite the horror expressed by my housemates when they found plastic bags full of it in the fridge.

At last, in desperation, I contacted the Forestry Commission and asked them if they knew of a wood in my area which held muntjac and in which I could work. Surprisingly, they suggested that I take a look at Rushbeds Wood, a 40-hectare holding near Brill, about 14 miles north-east of Oxford. Rushbeds was a semi-natural ancient woodland which they were doing nothing with at that time—possibly there were muntjac there. I would be free to use it, if it suited my purposes.

Driving out of Oxford in the zoology department van towards Rushbeds Wood on a cold winter's morning, I passed the newly restored tower of Magdalen College, gleaming golden in the sun. Then I drove through the wooded tunnel of Headington Hill before striking out into the quiet countryside beyond Stanton St John. I slowly approached the hill village of Brill overlooking the Vale of Aylesbury, and stopped by its old windmill to look out to the north. There below lay the wood, a lovely expanse of dark brown and grey branches gently linking with the larger lakeside woodlands of Wotton House. Would this be my new domain, a place where I could begin to unravel the intricate mysteries of muntjac ecology?

Rushbeds Wood was perfect. The abundance of fresh muntjac dung and footprints (or 'slots') allayed my fears that the little deer had not favoured the area. It was flat, modestly endowed with paths and rides, and apart from a dense blackthorn thicket at the western end, it was possible to walk anywhere in the wood. Virtually no one visited. It lay in quiet repose as it had done for all the centuries since it had been sliced off from the original great forest. There had been no disturbance of any kind for several decades, and the wood had a deliciously wild, unmanaged feeling that made me feel deeply relaxed and at home in its complex vegetation and its dark, overgrown tracks crossed with fallen trees and deep swampy puddles.

The work was hard. Amongst other things, I had to carry out a systematic, quantitative survey of the wood's vegetation in order to study muntjac habitat preferences. This work took two summers and one winter, laying out hundreds of temporary 5-metre square plots with bamboo poles and string, and then estimating by eye the cover of various species of herbs and shrubs within them. Trees had to be measured using a different, time-consuming technique. The concentration needed to extract these numbers from the living world was taxing and exhausting, and it seemed unnatural. After working on two or three plots, needing to rest my tired mind, I would lean back against a tree looking up at the sky through the wonderful wild mesh of branches, listening to the wood living its life as a vast breathing being. I became part of this being, with its swaying branches, its crisscrossing birdsongs, and its invisible muntjac carrying on with their strange lives all around me.

During these meditative moments there was a profoundly healing sense of Rushbeds Wood as an integrated living intelligence, a sense that expanded beyond the wood itself to include the living qualities of a wider world of the atmosphere, the oceans and the whole body of the turning world. Rushbeds Wood in these moments seemed to be quite clearly and obviously alive, to have its unique personality and communicative power. These periods of communion were intensely joyful and relaxing, and contrasted markedly with the stressful effort to reduce the wood to quantitative measurements in my multiplying field notebooks. I noticed with interest that the joyful sense of union would fade into the background of my consciousness as soon as data collection began. Gathering numbers was mind-numbing; being and breathing with Rushbeds Wood was liberating.

I had similar experiences whilst working at Whipsnade Zoo, where muntjac were free to wander almost anywhere within the spacious, park-like grounds. Here was a place where I could observe muntjac without the intervention of the dense, thickety vegetation of Rushbeds Wood, which afforded only fleeting glimpses of the muntjac as they crossed a clearing at dusk or dawn. The open, wooded lawns at Whipsnade made it easy to watch the little deer, many of whom I came to know as individuals. Once again, my brief was to collect numbers, this time about their movements and behaviour, so I would record on data sheets what the deer did and where they were every four minutes, for hours on end.

During my rest periods I would simply sit among the muntjac without collecting data at all. I particularly loved finding an animal that was chewing its cud. Sitting at a respectful distance, I would feel the intense, tranquil pleasure that seemed to emanate from the little animal as a bolus of food bulged along its oesophagus and into its mouth. I loved the half-closed eyes, the meditative tranquillity, and the delicious, warm, chamois-leathery sort of feeling that exuded from them like the aroma from a richly scented flower. It was as if a gentle yellow light emanated from them into the surroundings. My own animal body gleaned something of the ease and comfort with which they lived their lives, as though they were informing my senses with a kind of contentment I had not known before.

It is now more than 20 years since I did this work, and looking back I realise that I learnt as much, and possibly more, from the simple exposure of my own sensing organism to Rushbeds Wood and the muntjac than I did from the data collection and analysis that I was engaged in to gain my doctorate. Of course, analysing data and writing up the results were enjoyable pursuits in their own right that trained my rational mind and made it possible for me to become a card-carrying member of the scientific community. The science also allowed me to put together a fascinating and factually based account of the lives of the little deer that would have been impossible to achieve in any other way. But the learning that ultimately gave me the most valuable lessons about nature came from the unexpected qualities revealed to me by Rushbeds Wood and by the gently ruminating Whipsnade muntjac.

To my intense disappointment, there was no place for an exploration of these qualities in the fat doctoral thesis that I eventually submitted, for they were considered to be just my own subjective impressions. They were suitable for poetry perhaps, but did not belong to a way of doing science that wanted to banish me to a soulless world of bare facts devoid of inherent meaning. In an eloquent expression of this outlook, Bertrand Russell, the great 20th-century English philosopher, said that "Our origins, hopes and fears, our loves and beliefs are but the outcome of accidental collocations of atoms." In similar vein Jaques Monod, the much respected Nobel laureate in biochemistry, thought that the science that he practised required man to "wake to his total solitude, to his fundamental isolation", to "realise that, like a gypsy, he lives on the boundary of an alien world".

It was only when I came to work at Schumacher College, some three years later, that I encountered the notion that the major flaw with this perspective is the belief that the whole of nature, including the Earth and all her more-than-human inhabitants, is no more than a dead machine to be exploited as we wish for our own benefit, without let or hindrance. This idea, which has held centre court in the Western mind for about 400 years, has led us to wage an inadvertent war on nature, of gargantuan proportions. The casualties are mounting even as you read these words. Key indicators of planetary and social ill-health are growing exponentially fast, including species extinctions, water use, the damming of rivers, urban populations, the loss of fisheries, and average surface temperatures. It is a war that we cannot possibly win, as E. F. Schumacher so drily observed when he said that "Modern man talks of the battle with nature, forgetting that if he ever won the battle he would find himself on the losing side." We are living through a world-wide crisis of our own making: the crisis of 'global change'.

Many green thinkers agree that this mechanistic world-view has brought us to the brink of a catastrophe so great that our very civilisation is threatened, and that we urgently need to make peace with nature by re-discovering and embodying a world-view that reconnects us with a deep sense of participating in a cosmos suffused with intelligence, beauty, intrinsic value and profound meaning, as I had discovered at Rushbeds Wood. In this book we shall try to explore this participatory understanding using insights from Gaia theory, holistic science and deep ecology. In particular, we will ask to what extent it is possible to use recent scientific discoveries about the Earth to develop a deep reverence for our planet home so that we can then engage in actions consistent with this reverence, for science is a dangerous gift unless it can be brought into contact with the wisdom that resides in the sensual, intuitive and ethical aspects of our natures. As we shall see, it is only when these other ways of knowing complement our rational approach to the world that we can truly experience the living intelligence of nature.

Rediscovering Animism

The experiences of wholeness into which I had stumbled whilst living and working with muntjac were healing and full of significance, but my confidence in them had been almost totally undermined by the mechanistic views so eloquently articulated by Monod and Russell. I left the university with my doctorate, but also with a great deal of unease. Were Monod and Russell right, or was there anything of genuine value in the diverse life and intelligence that I had sensed in Rushbeds Wood and its inhabitants? And if what I had experienced was indeed real, could it ever become part of science?

For most non-Western cultures, such experiences of the living qualities of nature are a source of direct, reliable knowledge. For them, nature is truly alive, and every entity within it is endowed with agency, intelligence, and wisdom; qualities which in the West, when they are recognised at all, have commonly been referred to as 'soul'. For traditional cultures, rocks are considered to be the elders of the Earth; they are the keepers of the oldest memories and are sought out for their tranquil, wise counsel. High mountains are the abode of powerful beings, and are climbed only at the risk of gravely offending their more-than-human inhabitants. Forests are living entities, and must be consulted before a hunt by the shamans of the tribe, who have direct, intuitive connection with the great being of the forest. The American philosopher and cultural ecologist David Abram makes the point that many traditional peoples knew their natural surroundings as so intensely alive and intelligent, as so sensitive to one's presence, that one had to be careful not to offend or insult the very land itself. Thus, most indigenous cultures have known the Earth to be alive— a vast sentient presence honoured as a nurturing and sometimes harsh mother or grandmother. For such peoples, even the ground underfoot was a repository of divine power and intelligence.

These non-Western peoples espoused an *animistic* perspective, believing that the whole of nature is, in the profound words of 'geologian' Father Thomas Berry, *"a communion of subjects rather than a collection of objects"*. Animism has traditionally been considered backward and lacking in objective validity by Western scholars, but today philosophers, psychologists and scientists in our culture are beginning to realise that animistic peoples, far from being 'primitive', have been living a reality which holds many important insights for our own relationships with each other and with the Earth. One such insight is that animistic perception is archetypal, ancient, and primordial; that the human organism is inherently predisposed to seeing nature as alive and full of soul, and that we repress this fundamental mode of perception at the expense of our own health, and that of the natural world.

Psychologists involved in the study of child development recognise that children pass through an animistic phase in their early years, during which they relate to objects as if they had a character and as if they were alive—evidence consistent with my argument for the primacy of animism. But tragically, these same psychologists hold that this animistic phase is only appropriate to early childhood, and that one must help children to realise as quickly and painlessly as possible that they live in a dead world in which the only experiencing entities are other humans. However, not all psychologists subscribe to this view. James Hillman, a close student of Jung and the founder of Archetypal Psychology, suggests that animism is not, as is often believed, a projection of human feelings onto inanimate matter; but that the things of the world project upon *us* their own 'ideas and demands', that indeed any phenomenon has the capacity to come alive and to deeply inform us through our interaction with it, as long as we are free of an overly objectifying attitude. Hillman points to the danger of identifying interiority with only human subjective experience; a gaping construction site, for example, or a clear-cut mountainside, may communicate the genuine, objective suffering of the Earth, and one's sensing of this is not merely a dream-like symbol of some inner process which relates only to one's own private inner self.

This animistic perspective has a long and distinguished philosophical pedigree. For some eminent philosophers such as Spinoza and Leibniz, and more recently Alfred North Whitehead, it was inconceivable that sentience (subjective consciousness) could ever emerge or evolve from wholly insentient (objective, physical) matter, for to propose this would be to believe in a fundamental division or inconsistency within the very fabric of reality itself. Therefore each of these philosophers considered matter to be intrinsically sentient. The new animism that they espoused simply recognises that the material world around us has always been a dimension of sensation and feelings—albeit sensations that may be very different from our own—and that each entity must be treated with respect for its own kind of experience.

But if animism is indeed such an archetypal, primordial mode of perception, how did it come to be suppressed in such an effective and pervasive way in Western culture? What drove animism underground, and what have been the effects of its loss? The story of how animism vanished is complex, intricate, and difficult to fully unravel. Some theorists, such as Paul Shepard, suggest that the separation began with the widespread adoption of agriculture during the Neolithic (new stone age) period some 5,000 years ago. Shepard argues that agriculturalists developed a fearful attitude to undomesticated nature because their crops were continuously susceptible to pests, floods, droughts and other natural misfortunes, and because these early farmers had to expend a great deal of effort to prevent wild vegetation from taking over their fields and pastures. There is evidence that this fearful attitude was linked to the worship of wrathful masculine gods who were distant from nature and who had to be constantly placated in order to keep the more-than-human world under control. It seems to me that in the Neolithic era our spontaneous animistic sensibilities gave way to a dualistic animism, in which crops, fields and domesticated livestock had to be protected from the surrounding wilderness. According to author Roderick Fraser-Nash, the precursor of our word 'wilderness' comes from the concept of *wildeor,* from the 8th-century Beowulf epic connoting a mixture of 'will'—self-willed, uncontrollable nature and *deor*, meaning savage beast. Hence 'wilderness' is the place where uncontrollable dangerous beasts lurk, darkly threatening the agriculturalist's world.

Cultural ecologist David Abram holds that the advent of formal writing systems—and, in particular, the emergence and spread of the phonetic alphabet—was a major factor in the breakdown of the animistic experience. In his book *The Spell of the Sensuous* he demonstrates that phonetic reading involves a displacement of our instinctively animistic style of perception away from surrounding nature to the written word, such that the printed letters on the page begin to speak to us as vividly as trees, rivers, and mountains once spoke to our more indigenous ancestors. Writing and reading, according to Abram, involve a sublimated form of animism: while our indigenous forebears once participated, animistically, with animals, plants, and indeed every aspect of the expressive cosmos, we now participate exclusively with our own human-made signs and technologies.

Such a thesis helps explain why the richly animistic and nature-based polytheism of ancient Greece slowly transformed, in the 4th century BCE, to the more rational world-view of Plato and the philosophers. Plato himself was being educated in Athens precisely at the moment when the new alphabet was first included in the Athenian curriculum, so it was only natural that he would be one of the first to enact a new style of thought made possible by the phonetic alphabet. While the Homeric Greeks experienced surrounding nature to be filled with gods—they felt the presence of Zeus in a thunderstorm and Poseidon in the ocean waves—Plato articulated a new way of seeing and feeling, according to which the sensuous cosmos that we see around us was not the only potency in the world. For Plato, and the students in his academy, the perceived things that populate this world—the material things we see around us, which are subject to change, to growth and decay—are not the only reality. They are, rather, derivative copies of bodiless, eternal ideas that exist in some abstract realm. These archetypal ideas exist elsewhere, outside the body's world; the rational intellect alone has the capacity to gain access to that eternal domain beyond the stars.

Plato thus inaugurated the notion of an eternal heaven hidden beyond the material world, an ideal realm where the true source of things really exists. In the hands of later philosophers, and of the Christian Church, this notion led to an increasingly dualistic way of thinking, according to which everything genuinely meaningful and wondrous about the world was assumed to exist elsewhere, in some otherworldly dimension; while the sensuous, material world of nature was viewed as an illusory, derivative, and increasingly drab world, fallen away from its divine source.

But Plato himself was perhaps less of a dogmatic dualist than those who followed him, and may be best understood as a dualistic animist. In one of his richest writings, entitled the *Timaeus*, he articulated an idea that would have powerful repercussions during the European Renaissance almost two thousand years later. This was the notion that became known, in its Latin version, as the *anima mundi*—the 'soul of the world'. In his

Timaeus, Plato states that "This world is indeed a living being supplied with soul and intelligence . . . a single visible entity, containing all other living entities." Hence the world itself was considered to have a soul—the *anima mundi*—which had given birth to matter and then caused it to remain in ceaseless motion. *Anima mundi* was feminine, and permeated every aspect of the material universe.

In the *Timaeus* Plato also writes of a divine Creator who had enfolded the laws of mathematics and the beautiful symmetries of geometry into every aspect of the world. Although he also suggests in this work that every being was contained in and nourished by *anima mundi*, he nonetheless seems to insist on the primacy of the human intellect over emotion, the body, and the rest of the material world. According to philosopher Mary Midgley, for Plato the aim of human existence was to engage in intellectual enquiry into the laws governing the motions of the stars and planets, because the celestial realm was where the divine intellect was best displayed and comprehended. The Earth, however, was the realm furthest from the divine mind, and being full of imperfections, conflicts and contradictions, could be largely disregarded and to some extent disparaged, even though *anima mundi* was its creator and imposed upon it an overarching harmony that prevented a descent into total disorder.

In some of his dialogues Plato proposed a seamless interconnectedness of existence within a hierarchical ordering of the cosmos. The human soul was connected to the souls of animals and plants through the *anima mundi*, but failed humans reincarnated as animals, which were not worthy of much respect because they "came from men who had no use for philosophy". For Plato, the Earth was merely a convenient place from which to carry out the contemplation of celestial bodies, but any other habitable planet would have done just as well, since not much attention needed to be paid to what was after all merely the lowly abode of the body. Plato's most famous student was Aristotle, who, returning to a more sophisticated expression of the ancient non-dualistic animism, espoused a more immediate felt relationship with nature, in which each being was not the imperfect manifestation of a disembodied eternal idea, but displayed instead its own dynamic coming into being entirely from within itself.

A dualistic interpretation of Plato's ideas was later incorporated into Christianity, which before the Reformation espoused a peculiar kind of hierarchical animism in which man occupied a privileged position halfway between physical matter and the spiritual world. During the Middle Ages, the common folk who had no access to reading and writing were deeply animistic, and believed that nature was sacred, despite the efforts of the Church and its priests to impose the view that there were no spirits in trees, rocks, streams or forests. Instead, the priests attempted to convince the peasantry that these entities did not possess their own internal God-like powers, but rather that God had intended them merely as a sign of his own divine presence, which emanated from some disembodied, invisible domain far from the world of matter. But the animism of the common folk was resilient, and defied the efforts of medieval Christianity to stamp it out. As a result the Church simply adapted and compromised by taking over many ancient sacred places and by tolerating certain kinds of animistic practice. This peculiar and complex syncretism between animism and Christianity held sway for about 1600 years, until the birth of modern science.

The Scientific Revolution

This gradual sense of separation from nature in Western culture was greatly intensified during the scientific revolution which blossomed in the 16th and 17th centuries, in the wake of the Thirty Years War (1618–1648) that had decimated Europe in what had been a horrific conflict triggered by the break-up of the Church at the time of the Reformation in the 15th century. Plagues and famines had swept through Europe killing millions of people, and the war itself had generated massive destruction of property and loss of human life—one-third of the population of central Europe had been killed in it. The old, comfortable certainties that had held society together throughout the Middle Ages had broken down, and as the old world order collapsed under the pressures brought to bear on it by the new Protestantism, people felt intensely vulnerable and insecure. The old Church had christianised the ancient pagan religions, and still tolerated the animistic views of the majority of its congregation, but the Protestant revolution denied even this, and declared that God was detached from his physical creation, which was nothing more than a sinful, fallen realm that could be escaped upon death if one had worked hard enough to deserve a place in heaven. It was into this context that modern science was born. Its earliest practitioners and proponents, amongst them Bacon, Descartes and Galileo, were convinced that a new basis for certainty must be founded on reason rather than on a simple faith in established religious dogmas and what were seen as the superstitious beliefs of the common people.

Galileo (1564–1642) taught that one must ignore subjective sensory experiences if one wanted to learn anything useful about the world. Such experiences, such as empathy with a small deer chewing its cud, or awe at the diverse and elemental beauty of a forest in which one is wandering, were for Galileo unreliable and downright misleading. The English philosopher John Locke (1632–1704) gave the name 'secondary qualities' to such felt experiences, in order to emphasise their inferior, ostensibly derivative status relative to the primary qualities of size, shape and weight. Primary qualities—those which were rightly attributed to the objective, real world—were those aspects of things, and only those, that were amenable to quantitative measurement. Galileo believed that reliable knowledge resided in quantities, so nature had to be reduced to numbers if she was to yield her secrets and submit to the controlling influence of the human mind. For scientists, mathematics became the language for understanding and controlling nature. The justification for this was straightforward enough. After all, as philosopher Henri Bortoft points out, all right-minded, rational people agree on the correctness of mathematical propositions—no one disputes that all the angles in a triangle add up to 180 degrees. The new mathematical way of thinking was compelling precisely because it seemed to provide an indisputable, solid foundation on which science would build a new era of social stability based on the application of pure reason to every aspect of life.

Francis Bacon (1561–1626), like Galileo, was one of most important progenitors of the scientific revolution. He called for scientific researchers to "bind" and constrain nature using mechanical inventions so that she "could be forced out of her natural state and squeezed and moulded", and thereby "tortured" into revealing her secrets. According to Bacon, nature, once enslaved, "takes orders from man and works under his authority", and is thus put into bondage in order to expand human dominion over the physical universe.

The new science received a huge boost on the 10th November, 1619, at Nenberg on the banks of the Danube, when René Descartes (1596–1650) received a vision of the material world as a great machine. Descartes began to assert a fundamental distinction between matter, which he called *res extensa* (or 'extended stuff'), and mind, which he called *res cogitans* (or 'thinking stuff'). In essence, Descartes declared that the material world we see and sense around us was devoid of soul, and that it was nothing more than a dead, unfeeling machine which we could master and control through the exercise of our rational intellect. For him, the only non-mechanical entity in the universe, the only locus of subjectivity and soul, was the human psyche itself. Descartes taught that any entity could be completely understood by studying how its component parts worked in isolation—this was his famous reductionist methodology. His belief in mechanistic reductionism was so extreme that he urged his students to ignore the screams of vivisected animals, for such sounds were, after all, little more than the creakings and gratings of a complicated machine.

The work of the great English scientist, Isaac Newton (1642–1727), seemed to validate this emerging mechanistic world-view. Newton invented differential calculus—the mathematics of change—without which modern science would be impossible, and for which we owe him a great debt of gratitude. (In fact, Leibniz independently invented differential calculus at the same time.) Newton's equations stunned his contemporaries with their ability to precisely predict the trajectories of moving bodies such as cannon balls and orbiting planets, and seemed to provide final confirmation that the world was indeed no more than a vast machine whose behaviour could be precisely predicted and explained by means of quantification, reductionism and systematic experimentation. The new scientific method that these great thinkers established was thus entirely based on mathematical reason, and the key to its practice required the scientist to separate his mind from the rest of nature (which was considered to be an independently existing, objective reality) so that he could become an emotionally detached, strictly dispassionate instrument for the collection of data and for the observation of mechanical processes. Subjective impressions were eliminated, since these interfered with and invalidated the method and its results. Phenomena had to be investigated by means of carefully executed experiments in which all variables but the one under investigation were held constant. The results of these experiments were considered valid only if they could be replicated by other researchers, and if they could be used to make mathematical models predicting the future behaviour of the phenomena, in order to gain complete mastery and control over them.

The old religious certainties which had been so deeply brought into question with the break-up of the Catholic Church were now replaced by a new confidence in scientific materialism, which swept through the

Western world like intellectual wildfire, gathering momentum as more and more phenomena in nature fell under its sway, transforming the lives of millions of people all over the world. As mechanistic science grew in influence, the *anima mundi* faded from consciousness, so that now, some 400 years later, we have the dazzling technologies and scientific theories which are so much part of the cultural scene in the modern world; but we have lost contact with our deep animistic reverence for rocks, mountains, streams, rivers, and indeed for the whole of nature as a living intelligence. Yet if our awareness of the *anima mundi* is indeed an archetypal mode of perception, it can never be eradicated from the human psyche, even though it is possible to repress it within individuals or, it seems, within an entire culture. According to anthropologist Robert Lawlor, an awareness of the suppression of animistic consciousness is common to all indigenous tribal people today, who "believe that the spirit of their consciousness and way of life exists like a seed buried in the earth. The waves of European colonialism that destroyed the civilisations of North America, South America and Australia began a five-hundred-year dormancy period of the archaic consciousness. Its potencies disappeared into the earth."

Psychologists know only too well that what is repressed can haunt consciousness in the form of pathological behaviour and distorted perceptions. Is it any wonder then that our repression of *anima mundi* has come back to haunt us in the guise of a global crisis that is causing such massive destruction of wild nature and of traditional cultures? Sadly, by separating fact from value and quantity from quality, mechanistic science has inadvertently played its part in these disasters: the atom bomb, intensive agriculture, the ozone hole and climate change are examples of unintended but seriously damaging consequences of this overvaluation of the rational mind.

The crisis is at root one of perception; we no longer see the cosmos as alive, nor do we any longer recognise that we are inseparable from the whole of nature, and from our Earth as a living being. But there is hope, for as the crisis deepens, the call of *anima mundi* intensifies. More and more people are waking up to their deep connection to the intelligence of the cosmos, and are seeking to find ways of living that do not violate their rediscovered ecological sensibilities. It is as if the *anima mundi* is trying to express herself in our consciousness in ways which move beyond the dualistic animism of Plato and the otherworldly dualism of the old Church. In this time of crisis, we need only pay heed to our thorough embeddedness within the earthly web of life to feel the buried seed of *anima mundi* begin to stir and blossom in our minds and sensing bodies. As the seed breaks open, we see the wisdom in letting go of the objectivist assumptions of modern science, without abandoning the considerable achievements and benefits that it has undoubtedly brought us. This dawning awareness of the *anima mundi* in our times is in truth a reawakening of the old, non-dualistic animism that has been dormant for so long. It is a reassertion of our indigenous soul, and of the felt solidarity with earthly nature common to our indigenous, tribal ancestors. Our task now is to explore ways in which the new animism can be integrated into the very heart of Western culture. Holistic science is one possibility.

Holistic Science

There has always been a holistic, integrative strand in Western culture, espousing an animistic understanding, that ran alongside the reductionist scientific mainstream. One can even make a good case that the integrative and reductionist modes of consciousness are both innate to the human organism, and that they have manifested in different cultures in different ways at different times. Historian Donald Worster suggests that these two strands have been present in Western thought since the ancient Greeks, calling them the 'Arcadian' and the 'Imperial' strands of ecological thought. Philosopher Richard Tarnas refers to these two modes of perception as the 'empirical' and the 'archetypal', and novelist-scientist C. P. Snow spoke of the 'two cultures', science and the arts, which to his mind were almost impossible to reconcile. Some cultures held the balance more skilfully, and with more awareness than our own culture has done until now. For example, in Hindu philosophy one finds both tendencies expressed in the writings of various teachers, some of whom espoused a radical form of reductionism reminiscent of Democritus, the Greek philosopher who declared that everything is composed of indivisible units known as atoms (from the Greek *a-tom*, that which is indivisible).

Holistic science weaves together the empirical and the archetypal aspects of the mind so that they work together as equal partners in a quest that aims not at a complete understanding and mastery of nature, but rather that strives for genuine participation with nature. But how to clearly explain the approach of holistic science? When a prospective student first asked me for a coherent definition of holistic science, his question disturbed me, and for several days I could give no satisfactory response. Then it suddenly came to me that I could provide a useful answer by drawing on a profound insight from the great Swiss psychologist C. G. Jung, who spoke of four main psychological functions, or ways of knowing, common to all humanity, namely: intuition, sensing, thinking and feeling. Jung arranged these as two pairs of opposites, as follows (Figure 1):

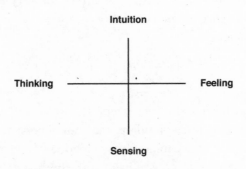

Figure 1: The Jungian 'Mandala'

In what I call the Jungian 'Mandala', sensation, or sensory experience, yields a direct apprehension of the things around us through the medium of our physical bodies. Thinking interprets what is there in a somewhat logical, rational manner; feeling grants a negative or positive valence to each encounter, and so helps to ascribe value to the phenomenon, and intuition yields a sense of its deeper meaning, as Jung says, "by way of unconscious contents and connections". Thinking interprets, feeling evaluates, whilst sensation and intuition are perceptive in that they make us aware of what is happening without interpretation or evaluation. Having treated hundreds of patients, Jung observed that each person has an innate conscious orientation towards one of the four functions, whilst the opposite function remains largely unconscious and undeveloped. The other two functions are only partially conscious, generally serving the dominant function as auxiliaries. Of course, this typology suffers from the limitations of all models, but Jung found it useful enough to say of it that it "produces compass points in the wilderness of human personality".

Mental and physical health in Jung's therapeutic approach required the

conscious development of the neglected function, together with an awareness of the four functions in oneself, so as to achieve a well-rounded
personality. Jung based his classification on ancient psychological systems
that also recognised the existence of four psychological types. Doctors in
the Middle Ages spoke of the four elements, air, fire, earth and water, whilst
ancient Greek medicine thought in terms of phlegmatic, sanguine, choleric
and melancholic personalities. Correspondences with the systems of other
cultures, such as that of the native North Americans or with the mandala
systems of India and the Himalayan Buddhists, strongly suggest that this
four-fold system has a certain cross-cultural archetypal resonance.

In our own culture, mainstream science is based principally on the deliberate cultivation of the thinking function, which is overly dominant not
only in science but in the culture as a whole. Feeling—the evaluative,
ethical function—is left out of science, so we might suspect that scientists
as practising professionals could greatly benefit from a solid grounding in
ethics as part of their training. In conventional science, sensation and intuition serve thinking as auxiliary functions. Sensation, the raw perception of
colours, sounds, tastes, touches and smells, is obviously essential for doing
science, for without it the world cannot register within our awareness.

However, the impulse in conventional science is to convert raw sensory
experience into numbers or abstractions as quickly as possible, often using
sophisticated scientific instruments to gather numerical information about
phenomena rather than use the naked senses directly. This mode of sensing
marginalises the phenomenon, and inhibits the possibility of the perception
of depth and intrinsic value in the thing being studied. I suffered from this
problem for many years, a condition that I call *quantificationitis*. A key
symptom was the inescapable compulsion to devise ways of measuring just
about everything. Dense, richly tangled vegetation caused me intense
suffering, for how on earth was I to measure all the multifarious ways in
which branches, stems and vines curled around each other like a multitude
of coiling snakes? How to measure all the subtle colours and leaf shapes
that delighted the senses, but boggled the intellect?

Thinking, the function most valued by our culture, is used in conventional science to devise experiments and to construct mathematically
coherent theories and accounts of how the world works. The predominant
style of thinking used in conventional science is reductionism, in which, as
Descartes taught, one attempts to gain complete understanding and
mastery of a phenomenon by breaking it down into its component parts.

Once the behaviour of each part is known, the reductionist approach teaches us that the behaviour of the whole will become intelligible as the summed effect of the individual components. Reductionism is built on a broader set of important assumptions, such as that objects matter much more than the relationships between them, that the world is ordered hierarchically, that knowledge can be objective, and that the knowing intellect can wholly detach itself from the material world in order to attain a purely objective, 'God's eye' view of any given phenomenon.

Reductionism works very well if we want to design things like cars and computers, but its success is more limited in areas such as biology, ecology or in the realm of human social life where complex, non-linear interactions are the norm. In these areas we need to apply a different style of thinking which builds on and incorporates reductionism whilst moving beyond it. Physicist Fritjof Capra points out that this new approach, which he calls 'systems thinking', involves shifting our focus from objects to processes and relationships, from hierarchies to networks and from objective knowledge to contextual knowledge.

What does this mean? A key insight in systems thinking is that we can understand a great deal more about a system if we focus on the patterns of relationship between the parts rather than on the parts themselves as isolated entities. We come to realise that the properties of the parts depend upon how they relate to each other and to the larger whole that they help to constitute. We also come to understand that there are no fundamental building blocks, such as subatomic particles, at the base of a hierarchical ordering of nature, but that nature self-organises into multi-levelled sets of networks within networks, such as cells within tissues within organs within organisms within ecosystems within Gaia, and that no one level is fundamental. Finally, we realise, as did Werner Heisenberg, one of the great physicists of the early part of the last century, that *"what we observe is not nature herself, but nature exposed to our method of questioning"*; that our knowledge depends on how we interact with the world.

When we focus on relationships between the parts rather than on the parts in isolation we very quickly encounter the principle of *emergence,* in which surprising properties appear at the level of the whole that cannot be understood through focusing on the parts alone. A common adage which expresses this insight is that the 'whole is greater than the sum of the parts'. Emergence is widespread in nature. Good examples abound in the realm of social insects, where interactions between individual bees, ants or termites, each obeying simple rules for when to engage in activities such as searching for food, tending the brood or nest building, give rise to complex behaviours at the level of the colony that defy reductionist explanation. Individual ants of the genus *Leptothorax* are active or inactive according to a basically erratic schedule, but when they interact together at the right density the colony throbs with collective rhythmic activity. The complex, baroque labyrinths that constitute the interior of termite mounds are built by individual termites, each depositing little dollops of mud saturated with a pheromone, or chemical signal, that gradually evaporates away into the surrounding air. Termites are attracted to the pheromone, and deposit new dollops of mud wherever they encounter it at high concentrations. Computer models of this process show that an initially random pattern of mud dollops soon turns into an emergent, regular array of pillars and columns which look remarkably similar to the insides of real termite mounds. None of the ants or termites had the 'blueprint' for the right behaviour or mound structure—they emerged out of the interactions and relationships between the members of the insect social group.

Systems thinking involves stepping away from the notion that it is possible to predict and control nature, at least in anything but very limited ways. This insight has come in part from the practical experience of inventors and physiologists who have found it hard to predict the outcome of interactions in even simple non-linear systems, because slight changes in initial conditions mean that after just a short time the system can easily find itself in a totally different state—a bat flapping its wings in England could in theory lead to a tropical storm in the Amazon. If we cannot predict the exact nature of emergent properties, and if small changes can have unforeseeable and potentially dramatic outcomes, we have to accept the possibly uncomfortable conclusion that nature is inherently unpredictable and uncontrollable. Indeed, systems thinking suggests that the metaphor of control is the wrong basis on which to build a fruitful relationship with nature—participation is clearly more appropriate, and is in fact the only available option. In order to participate fully and properly we need to use quantitative methods appropriately, but we also need to develop a deep, intuitive sensitivity to the qualities of things. So what is intuition, and how can it be developed in the practice of holistic science?

Both conventional and holistic sciences are utterly dependent on intuition, for without it there would be no raw insights for the rational mind to work on and develop. Aristotle provides a classic example of how intuition works in science through what he called "direct induction", when he talks about the realisation that the light of the moon does not come from within the moon itself, but is in fact the reflected light of the sun. The dawning of this insight in him typifies the way in which intuition suddenly presents consciousness with a new way of seeing, often after the thinking mind has activated the unconscious through a concentrated focusing of attention on a phenomenon or on a given problem. Despite its importance, conventional science makes no effort to cultivate intuition as part of its methodology—it is seldom discussed, and its occurrence is mostly left to chance.

In holistic science there is an attempt to cultivate intuition as much as thinking, sensation and feeling. In cultivating intuition, many radical holistic scientists use a methodology which has been largely attributed to the German poet and scientist Johann Wolfgang von Goethe (1749–1832), but which in fact can be traced back several centuries before him to Ficino and Paracelsus, and before them to the Hermetic tradition. In this method, careful attention is paid to the phenomenon being studied through a process of *active looking*, without attempting to reduce the experience to quantities or explanations. For Henri Bortoft, active looking involves the "redeployment of attention into sense perception and away from the verbal-intellectual mind". In this way of seeing, one makes an effort to notice the specific details of the thing in all their particularities as they appear to the senses. If this works as it should, one can experience the suspension of one's preconceived notions and habitual responses about the thing being perceived, so that its exact sensorial qualities enliven and deepen perception. This allows the phenomenon, as Bortoft says, "to coin itself into thought", and "to induce itself in the thinking mind as an idea". One has the intuitive perception of the thing as a presence *within oneself*, and not as an object *outside* one's own being. This sense of deep related-ness to the object transforms consciousness into a means for holistic perception through which we are able to apprehend the intrinsic qualities of things. The methodology develops what can be called 'non-informa-tional perception', in contrast to the conventional approach, which stresses perception in the service of information gathering. Non-informa-tional perception is a subtle mode of perception that brings a sense of wholeness, whilst informational perception, although coarser, gives access to the quantitative dimension of reality that makes measurement possible.

Goethe paid careful attention to how intuitive insights dawned on him in the very midst of careful observation, and was very interested in how Galileo used his own intuitive perceptions of swinging pendulums to develop an understanding of the behaviour of falling bodies in general. For Goethian scientist Margaret Colquhoun, the process of doing Goethian science involves four steps. First is *intuitive perception*, which occurs spontaneously when one encounters a phenomenon without preconceptions through active looking. Next comes *exact sensing*, which involves a careful and precise examination of the parts of the phenomenon, such as, in a plant, the shapes and colours of leaves, the pattern of their arrangement on the stem, where there are buds and hairs, and so on, in great detail, to the extent that one almost deliberately allows oneself to lose sight of the whole. Conventional science also does this extremely well, but moves on to make reasoned, testable hypotheses and theories about the underlying mechanisms that could have brought the plant into being. Goethe asks us to suspend the urge to theorise, and to enter as fully as we can into the experience of sensing the phenomenon before our gaze.

This intention bears fruit in the next stage, *exact sensorial fantasy*, in which we close our eyes and allow the details we so carefully observed in the previous stage to flow together in our imagination as a coherent unfolding of life and form. As we visualise the plant sprouting from seed, growing, flowering, seeding and dying we may be fortunate enough to enter into the next stage, *seeing in beholding*, where we are given a revelation of the inner being of the plant and glimpse its holistic sacred quality. The final stage has been called *being at one with*, in which we have returned to a state of what Margaret Colquhoun calls 'intuitive precognition', in which we commune with the unbroken wholeness of the phenomenon, realising that each worldly thing is a manifestation of a single immanent loving creative energy.

This approach is best practised communally, so that it becomes possible to discriminate between what is common to the perceptions of a group of investigators and what could be merely idiosyncratic fantasy and projection. A similar approach, although not based on the Goethian methodology outlined above, has been used with great success by the animal welfare scientist Françoise Wemelsfelder, who has found that people's subjective assessments of the physical and mental states of farm animals are highly correlated, and are a very good indicator of the overall health of the animals.

Active Looking

Hold a small stone comfortably in your hand, and keep it at the same orientation for the duration of the exercise. Relax, and let go of any determination to achieve a result.

Now look carefully at the parts of the stone's surface. Pay very careful attention to all the subtle changes in colour and texture, to any scratches or marks, to any dimples or hollows. Do this for 30 seconds to a minute, and then close your eyes. Now, for about a minute, visualise what you have just seen as clearly as possible with as much detail as you can. Now let the image go, and just do nothing for a few seconds.

Open your eyes and look at the stone as a whole, without focusing on the details at all. Allow yourself to take in the stone as an entirety, as a single unified phenomenon. Allow the wholeness of the stone to waft into your being without asking yourself what the whole is or how you can actually see it. As before, do this for 30 seconds to a minute. Now close your eyes and in your mind's eye see the stone as a whole for about a minute. Then let the image go.

Repeat this cycle for about fifteen minutes, and then enter into a quiet period of reflection. Were there any differences between the two ways of looking?

Intuitive, holistic perceptions of wholeness naturally connect us with Jung's feeling function, that is, with the domain of ethics. Ethics, simply put, is the ability to decide whether a thing is right or wrong, whether it is good or not. Conventional science ignores ethics, leaving it to society to decide how to use the fruits of scientific research in the world at large. However, in holistic science we realise that perceptions of wholeness arrived at through active looking are inseparable from a deep sensitivity to the intrinsic value in the being or entity we are interacting with, making it very difficult for us to do anything that would harm or disturb the 'inner necessity and truth' of that being. For holistic scientists, many forms of genetic engineering are ethically unacceptable because intrinsic natures are violated when alien genes are transferred from one being to another. These perceptions of intrinsic value have practical consequences—they draw holistic scientists into public debates that concern their areas of research.

Holistic science is thus about reuniting fact and value in ways that enable our culture to explore new possibilities of living harmoniously with the Earth. This work involves integrating an animistic relationship with the Earth back into Western culture; clearly a difficult challenge, since the objectivist view opposes any notion that the universe is alive, creative and intelligent. This is where holistic science could be of great value by showing how it is possible to embed animistic insights into an expanded science that combines qualities with quantities whilst taking into account the ethical dimensions of participating in a living cosmos.

Language is a key aspect of this work, and so in this book we will experiment with a new kind of narrative that tries to explore the dynamics of our living Earth in a way that uncovers the beauty, way of being and vitality of her processes without falling into the dull mechanistic style that so dominates modern science, and that so deadens the world with its desic-cating touch. This language is still struggling to be born, and so I ask you to be patient with my faltering efforts to articulate it in various ways.

In searching for this language, we need to fearlessly adopt what James Hillman, the founder of Archetypal Psychology, has called "personifying", which he defines as the *"spontaneous experiencing, envisioning and speaking of the configurations of existence as psychic presences"*. We need to allow ourselves to be open to the subjective agency at the heart of every 'thing' in the world so that we can speak and act appropriately in their presence and on their behalf. We must keep alive and nurture a sense of the 'otherness' of whatever phenomenon we might be considering, allowing a strange kind of intimacy to develop in which the urge to control is replaced by a quickening awe at the astonishing intelligence that lies at the heart of all things. We must oppose the tendency of conventional science to de-personalise the world and hence to control it. We must oppose its desire to scrape away all subjectivity and to make us feel that science is value neutral—for if the world truly feels, then we cannot look at the world as outsiders—we are related to it and embedded in it, and the ethical dimension is there with us right from the start. This way of speaking recognises that for our sensing, feeling and intuition the whole of nature is a vast encompassing *being*, whereas for our thinking it is also a complex, interconnected *system*. Thus, holistic science attempts to develop a language which talks about the being or *life* of things—of their felt, existential quality, without alienating the rational mind.

Hillman points out that in personifying we recognise that the world at large is a communion of *persons* in the widest more-than-human sense, that subjectivity is not just the prerogative only of human beings. The emphasis on an expanded notion of personality is a key aspect of the new animism, which, in the words of philosopher Graham Harvey, is concerned with "taking seriously the intimations that the term 'person' applies not only to humans and human-like beings . . . but to a far wider community", and with asking how we are to treat or act toward such persons. Native peoples around the world are our best teachers in this regard. A particularly accessible example comes from the insights that anthropologist Irving Hallowell gleaned whilst living with the Ojibwe people of southern central Canada, who speak of 'bird people', 'bear people' and even, occasionally, 'rock people', because these are for them all subjects embedded in a wider world of complex participatory relationships. For Hillman there is an intimate connection between personifying and loving. In his words: "Loving is a way of knowing, and for loving to know it must personify. Personifying is thus a way of knowing, especially knowing what is in the invisible, hidden in the heart". In science, Descartes' fundamental division between living human subjects and dead external objects has seen to it that personifying (and the loving that accompanies it) are considered nothing more than mere projection and 'fantasy'. But today we can now realize that it was Descartes who was projecting, and that his fundamental division of mind from matter was itself a great fantasy—a chimera that we need only dissolve in order to find our true home in the great psyche of the world.

If you have been trained, as I have been, to see the world as a machine and to see yourself as not much more than a thinking, emotionally detached data-collecting robot, then to personify the world in this way takes a great deal of courage. As I wrote this book, the unspoken scientific taboo against speaking of the world as a psyche exerted its influence on me and tried its best to make me write nothing more than straightforward popular science. A strange vulnerability, an insecurity, sometimes plagued me as I attempted to speak of the Earth and of the living beings that inhabit her not merely as objects, but as subjects, as feelingful beings, but in the end a still, small voice persuaded me of the urgency of the task. In quiet moments in my study, or outdoors, this deeper voice convinced me that the prospects are bleak unless we can once again relate to the Earth not as a thing or as a machine, but as a strange creature that improvises its own unfolding in the cosmos through the ongoing creativity of evolution and self-transformation. As you notice the tension between these different voices, I ask you to remember the difficulty of the task and to consider yourself a conspirator in the effort to find a new language for breathing life back into our experience of the Earth, who for the last 400 years has been treated as if she were a dead lump of rock with a few insignificant and rather irksome life forms and traditional cultures clinging to her ragged surface. And now the title of this book reveals its double meaning, for 'animate' is both an adjective and a verb. The adjective tells us that the Earth is animate—*alive*; the verb urges us to find ways of speaking and acting that allow us to consciously re-animate the Earth so that we bring her back to life as a sensitive and sentient Being—even, if you will, as a person in the widest and wildest sense of the word. It is time to rediscover Gaia, for Gaia is Earth personified.

Chapter 2

Encountering Gaia

Gaia and the Ancient Greeks

For millennia, traditional peoples all over the world have believed in an Earth mother who bestows life and receives the dead into her rich soil. The ancient Greeks called her Gaia, the earthly presence of *anima mundi*, the vast and mysterious primordial intelligence that steadily gives birth to all that exists, the great nourishing subjectivity—at once both spiritual and material—that sustains all that is. Gaia inhabited the subterranean caves at Delphi, Athens and Aegae and spoke directly to priestesses intoxicated by the vapours exhaled from deep vents in the Earth mother's womb. Hesiod (around 700BCE) tells us that "Gaia was born from the primordial chaos, vast and dark", and that she gave birth to Uranus the sky with its multitudes of stars and to Pontus, the seas and oceans. Gaia, the "deep-breasted Earth", then mated with her child Uranus, and gave birth to the gods and the Titans, who eventually engendered all the Earth's living beings, including our own species. This is part of Hesiod's homage to Gaia:

Gaia, mother of all,
the foundation, the oldest one,
I shall sing to Earth.

She feeds everything in the world.

Whoever you are,
whether you move upon her sacred ground,
or whether you go along the paths of the sea,
you that fly, it is she who nourishes you from her treasure store.

In fact, long before Hesiod and Homer, Gaia was considered to be the most powerful of all deities, far more important even than Zeus and his Olympian pantheon. Here is how Charlene Spretnak re-tells the pre-Homeric myth of Gaia:

From the eternal void, Gaia danced forth and rolled herself into a spinning ball. She moulded mountains along her spine, valleys in the hollows of her flesh. A rhythm of hills and stretching plains followed her contours. From her warm moisture She bore a flow of gentle rain that fed her surface and bore life.

We have already seen that the experience of Gaia as a living presence was gradually buried in the West under the accumulating sediments of other-worldly religiosity and of an exceedingly dualistic science, which saw the Earth as no more than a dead machine obedient to the blind laws of physics and chemistry. Nonetheless, despite the almost total oblivion to which the awareness of Gaia was consigned, faint traces of her presence can still be found in words very much used today, such as *ge*-ology, *ge*-ometry and *ge*-ography, which incorporate, unbeknownst to almost everyone, the seed-syllable *Ge*—which is none other than the ancient and original form of the name Gaia.

I think it reasonable to propose a simple hypothesis—that throughout Gaia's almost 4,000-year long period of exile, Gaian awareness has continued to appear in the minds of certain receptive individuals via the gateways of intuition, sensing, feeling and thinking; the four cardinal points of Jung's mandala of the psyche, and that this is happening with ever more urgency and intensity in our own time of profound ecological crisis. We will now look at how Gaian awareness has manifested through each of the four functions (or ways of knowing) in certain key individuals.

Intuition: Aldo Leopold and the Dying Wolf

Intuitive experience of Gaia can break thorough into consciousness quite unexpectedly, sometimes so powerfully that our outlook on life can be permanently changed. The famous American ecologist and wildlife manager, Aldo Leopold (1886–1948), provides a striking example in his book *A Sand County Almanac.*

Leopold is one of the fathers of the modern ecology movement, inspiring seminal activists such as Rachel Carson, and yet he began his career as a staunch believer in the reductionist approach, and was one of the founders of the science of wildlife management. Leopold adhered to the unquestioning belief that humans are superior to the rest of creation, and thought it morally justifiable to manipulate nature to maximise human welfare. He supported a US government policy to eradicate the wolf from the United States using scientific methods, the justification for this intervention being that wolves competed with sport hunters for deer, so that fewer wolves would mean more deer for the hunters.

Leopold held this view strongly for many years until one morning he was out with some friends on a walk in the mountains of New Mexico. Being hunters, they carried rifles with them in case they had an opportunity to kill some wolves. At lunchtime they sat down at the edge of a cliff overlooking a turbulent river. Soon, they saw what appeared to be a deer fording the torrent, but they quickly realised that it was one of a pack of wolves. Leopold and his friends took up their rifles and began to shoot into the pack, but with little accuracy. Eventually, an old wolf was hit, and Leopold rushed down the steep slope in exhilaration. But what he found there was something utterly unexpected, something strange and wild, something he'd never experienced before. What met him ignited a part of his own soul that had been dormant until that moment; he saw a fierce green fire dying in the old wolf's eyes. He writes that:

> . . . there was something new to me in those eyes, something known only to her and to the mountain. I thought that because fewer wolves meant more deer, that no wolves would mean hunter's paradise. But after seeing the green fire die, I sensed that neither the wolf, nor the mountain agreed with such a view.

Perhaps it is possible to understand the notion of a wolf disagreeing with such a view. After all, the dying wolf was a fellow mammal with whom

Leopold could feel a certain affinity. But how could a lifeless, inert mountain possibly agree or disagree with anything? What could Leopold have meant by that? What had he experienced at that pivotal moment? Clearly he used the word 'mountain' as shorthand for the wild ecosystem in which the incident took place, for the ecosystem as an entirety, as a living presence with its deer, its wolves and other animals, its clouds, soils and streams. For the first time in his life, Leopold felt completely at one with this wide ecological reality. He felt that it had a power to communicate a sacred magnificence. He felt that it had its own life, its own intelligence, its own history, its own trajectory into the future as a living personality. He experienced the ecosystem as a great being, dignified and valuable in itself. It must have been a moment of tremendous liberation and expansion of consciousness, of joy and energy—a truly spiritual or religious experience. His narrow, manipulative wildlife manager's mind fell away for good. The attitude that saw nature as a dead machine, as there solely for human use, vanished. Leopold had been *Gaia'ed*. He had recognised the existence of an active agency far greater than himself in the great wild world around him; in the rocks, the air, the birds, the sun, the microbes in the soil and in every speck of matter. He understood them all now as powers that had witnessed what he had done with profound disapproval.

Notice that the experience was not looked for, expected or contrived—it happened spontaneously. Something in the dying eyes of the wolf reached beyond Leopold's training and triggered a recognition of *where he truly was*. From that moment on Leopold saw the world differently, and as he strove—over the course of years—to find a voice appropriate to this way of seeing, he eventually wrote a key essay about his Land Ethic, in which he stated that humans are not a superior species with the right to manage and control the rest of nature, but rather that we are just "plain members of the biotic community". In this justifiably famous essay, he also penned his famous dictum: "A thing is right when it tends to preserve the integrity, stability and beauty of the biotic community. It is wrong when it tends otherwise."

The experience imprinted itself so powerfully in Leopold's psyche that he wrote the following passage combining rational insights about the Earth as a living being with profound intuitions and ethical guidelines about how we should live within such a world:

It is at least not impossible to regard the earth's parts—soil, mountains, rivers, atmosphere etc,—as organs or parts of organs of a coordinated whole, each part with its definite function. And if we could see this whole, as a whole, through a great period of time, we might perceive not only organs with coordinated functions, but possibly also that process of consumption as replacement which in biology we call metabolism, or growth. In such a case we would have all the visible attributes of a living thing, which we do not realise to be such because it is too big, and its life processes too slow. And there would also follow that invisible attribute—a soul or consciousness—which many philosophers of all ages ascribe to living things and aggregates thereof, including the 'dead' earth. Possibly in our intuitive perceptions, which may be truer than our science and less impeded by words than our philosophies, we realize the indivisibility of the earth—its soils, mountains, rivers, forests, climate, plants and animals—and respect it collectively not only as useful servant but as a living being, vastly less alive than ourselves, but vastly greater than ourselves in time and space. Philosophy, then, suggests one reason why we cannot destroy the earth with moral impunity; namely that the 'dead' earth is an organism possessing a certain kind and degree of life, which we intuitively respect as such.

Anima mundi had moved Leopold to the core of his being, and had triggered in him a holistic mode of perception in which his intuition, sensing, thinking and feeling functioned as a unified whole. Leopold had come home.

Finding Your Gaia Place

One of the best things you can do to promote your own mental health is to find a special place outside where you can go on a regular basis to connect with the animate Earth. If you live in a city, this may be your own back garden or yard, but if you live in the countryside you will almost certainly have access to a variety of inspiring places in your immediate surroundings. Wherever you are, your task is to search for a place where you can spend time exploring and deepening your relationship to the great living being that is our planet.

Make sure that you allow yourself to be guided by your sensing, feeling and intuition when you are looking for your special place—let your thinking take the back seat. You'll know that you've found the right place if it provokes a profound sense of pleasure in you (perhaps even a feeling of overwhelming beauty), if your senses tingle with amazement at its sheer loveliness. Pay attention to how the place works on your feelings. Choose a place that evokes an easy comfort in you, and notice how the place speaks to your intuition—is there a numinous 'aura' that connects you to the psyche of the place, to a sense of its animate presence? Lastly, let your thinking mind consider the logistical viability of the place—is it too far for regular visits, will it offer you enough privacy and quiet, what clothing and footwear and other outdoor gear will you need to make your visits peaceful and comfortable?

Develop a rapport with your place by visiting it regularly—allow it to communicate its subtle messages of colour, scent, taste, touch and sound. As you let yourself be known by your place, learn to converse with it, gleaning its subtle meanings much as you would enjoy a conversation with a close friend.

It might help to have several Gaia places, some less wild perhaps, closer to home, and others further out in wilder country for extended visits and overnight communion under the sparkling light of the stars.

Sensing: David Abram, Phenomenology, and the San People of the Kalahari

Perhaps the most eloquent modern exponent of how Gaia can become present to us through our senses is the philosopher David Abram. Abram's fascination with sensory perception stemmed from his many years as a professional sleight-of-hand magician; perception, he explains, is the primary medium of the magician's art. In his mid-twenties, he journeyed as an itinerant magician throughout rural areas in south-east Asia, meeting and learning from the traditional indigenous magicians, or shamans, who practised their craft in those village lands. Initially focused on the uses of magic in curing and folk-medicine, Abram was startled to discover that the traditional medicine-persons he lived with viewed their ability to heal other persons as only a secondary skill. Their power to heal was considered by them a by-product of their much more primary role, that of their function as *intermediaries* between the human community and the more-than-human realm of animals, plants, and earthly elements within which the human community was embedded. In the course of living with these practitioners, Abram began to sense the diverse life of the surrounding nature with an intensity he had never experienced in his home country, the United States.

One day, climbing on an island in Indonesia, Abram was unexpectedly trapped in a cave by the first torrential downpour of the monsoon season. As the runoff from the cliffs above gathered into a solid waterfall, sealing off the entire entrance to the cave, a small spider weaving its delicate web across the cave's entrance caught his attention. Watching the subtlety of the spider's spiralling movements as she set and tested the various parts of her web, Abram soon caught sight of another spider weaving its own web overlapping the first. This led him to adjust the focus of his eyes, whereupon he abruptly discovered that there were numerous other spiders spinning their spiral structures at various distances from his face as he watched, dazzled. The intricate activity of the spiders drew him deeper into a trance; he soon found that he could no longer hear the roar of the cascading torrent just behind the expanding webs. His senses transfixed, Abram began to feel that he was witnessing the universe itself being born, galaxy upon galaxy taking shape before his eyes.

When he was awakened by the sunlight streaming into the cave the next morning, Abram could find no trace of the spiders, nor their webs. But as he climbed down from the cave, he discovered that he was no longer able to see any aspect of the world as an inert or inanimate presence: even the rocks and the cliffs were shimmering with life. Like Leopold, Abram had been *Gaia'ed*.

In his subsequent writings, Abram began to describe the event of perception as a deeply interactive, participatory encounter—as a kind of non-verbal conversation between the perceiver and that which he or she perceives. Indeed, in a seminal essay published in *The Ecologist* in 1985, entitled 'The Perceptual Implications of Gaia', Abram points out that sensory perception could be recognised as a wordless communication between the encompassing sentience of Gaia and one's own individual awareness—for if soils, plants, animals, atmosphere and water are not just a random collection of passive objects and determinate, mechanical processes, but are in truth living, sentient entities, then every instance of perception conveys something to us about the state of that greater being in which we are embedded.

According to Abram, the bifurcation of mind from matter in the modern world has precipitated an extreme dissociation of our conscious, thinking selves from our bodies; in his writings he seeks to draw readers back to the simple experience of their own corporeality, coaxing them to notice the ongoing, improvisational way that their animal senses spontaneously respond to the sensuous surroundings. His writings carry forward the tradition of 'phenomenology', a branch of philosophy that studies our direct, pre-conceptual experience of the world (the way things reveal themselves to us in their felt immediacy, prior to all our theorising and categorising). Abram draws especially on the work of the great French phenomenologist Maurice Merleau-Ponty. But Abram takes up phenomenology only in order to transform it and bring it into new terrain, deploying its methods to elucidate our felt relationship with the rest of nature. For, as he writes,

> The recuperation of the incarnate, sensorial dimension of experience brings with it a recuperation of the living landscape in which we are corporeally embedded. As we return to our senses, we gradually discover our sensory perceptions to be simply our part of a vast, interpenetrating webwork of perceptions and sensations borne by countless other bodies—supported, that is, not just by ourselves, but by icy streams tumbling down granitic slopes, by owl wings, and lichens, and by the unseen, imperturbable wind. . . .

Abram takes up Merleau-Ponty's notion of the collective "flesh of the world" to speak of this vast, planetary tissue of sensations and interdependent perceptions in which our own lives (like those of the trees, the crows, and the spiders) are embedded. The term "flesh" provides Abram with a way of speaking of reality as a fabric woven of experience—and hence of the material world as thoroughly animate and alive. While his work draws multiple insights from the natural sciences, the notion of "the flesh of the world" offers Abram a way to describe the earthly biosphere not as it is conceived "by an abstract and objectifying science, not that complex assemblage of planetary mechanisms being mapped and measured by our remote-sensing satellites, but rather the biosphere as it is experienced and lived from within by the intelligent body—by the attentive human organism who is entirely a part of the world that he, or she, experiences."

According to this way of thinking, perception is never a unilateral relation between a pure subject and a pure object, but is rather a reciprocal encounter between divergent aspects of the common flesh of the world. Abram calls attention to the obvious but easily overlooked fact that the hand, with which we explore the tactile surfaces of the world, is itself a tangible, tactile being, and hence is entirely a part of the tactile field that it explores. Similarly our eyes, with which we gaze out at the visible world, are themselves visible. With their shiny surfaces and their brown or blue hues, the eyes are included within the visible landscape that they see. Our sensing and sentient body—with its own sounds, smells, and tastes—is entirely a part of the sensuous landscape it perceives. Hence, to touch the coarse skin of an oak tree with one's fingers is also, at the same moment, to experience one's own tactility, to feel oneself touched *by* the tree. Similarly, to gaze out at a forested hillside is also to feel one's own visibility, and so to feel oneself exposed to that hillside—to feel oneself *seen by* those trees. Whenever we experience ourselves not as disembodied minds, but as the very palpable, sensitive, and sentient organisms that we are, we cannot help but notice this curious reciprocity in our sensory experience: to perceive the world is also to feel the world perceiving us.

Abram's analysis of the mutual, participatory nature of sensory perception goes a long way toward helping us understand the near universality of animistic experience among indigenous, oral peoples. Here is an anthropologist Richard Nelson's brief description of the Koyukon tribe of central Alaska:

> Traditional Koyukon people live in a world that watches, in a forest of eyes. A person moving through nature—however wild, remote, even desolate the place may be—is never truly alone. The surroundings are aware, sensate, personified. They feel. They can be offended. And they must, at every moment, be treated with the proper respect.

From the opposite side of the planet, the great South African writer and explorer Laurens van der Post reports a striking example of the same mode of perception among the San people of the Kalahari desert. Far from the campfire an elder is instructing a younger member of his tribe:

> You may think that deep in the darkness and the density of the bush you are alone and unobserved, but that, Little Cousin, would be an illusion of the most dangerous kind. One is never alone in the bush. One is never unobserved. One is always

known. It is true there are many parts of the bush where no human eye might be able to penetrate but there is always, like some spy of God himself, an eye upon you, even if it is only the eye of some animal, bird, reptile or little insect. . . . And besides the eyes—and do not underrate them—there are the tendrils of the plants, the grasses, the leaves of the trees and the roots of all growing things, which lead the warmth of the sun deep down into the darkest and coldest recesses of the earth, to quicken them with new life. They too shake with the shock of our feet and vibrate to the measure of our tread and I am certain that they have their own ways of registering what we bring or take from the life for which they are a home. Often as I have seen how a blade of grass will suddenly shiver on a windless day at my approach or the leaves of trees tremble, I have thought that they too must have a heart beating within them and that my coming has quickened their pulse with apprehension until I can note the alarm vibrating at their delicate wrists and their high, translucid temples.

Indeed, the discourse of virtually all oral, indigenous peoples supports Abram's claim that "in the untamed world of direct sensory experience *no* phenomenon presents itself as utterly passive or inert. To the sensing body *all* phenomena are animate, actively soliciting the participation of our senses, or else withdrawing from our focus and repelling our involvement. Things disclose themselves to our immediate perception as vectors, as styles of unfolding—not as finished chunks of matter given once and for all, but as dynamic ways of engaging the senses and modulating the body. Each thing, each phenomenon has the power to reach and to influence us. Every phenomenon, in other words, is potentially expressive." Everything speaks.

According to Abram, then, our direct, spontaneous perception of the world is inherently animistic; only as perception came to be mediated by various technologies was this spontaneous and instinctive experience transformed in a way that robbed the perceived things of their felt vitality, draining them of their inexhaustible otherness and mystery. Abram holds that a genuine environmental ethic is not likely to emerge through the logical elucidation of new philosophical principles and legislative strictures, but rather "through a renewed attentiveness to this perceptual dimension that underlies all our logics, through a rejuvenation of our carnal, sensorial empathy with the living land that sustains us".

Modern science itself, for all its cool detachment, remains rooted in the soil of our direct sensorial experience. In the words of Merleau-Ponty,

"The whole universe of science is built upon the world as directly experienced, and if we want to subject science itself to rigorous scrutiny and arrive at a precise assessment of its meaning and scope, we must begin by reawakening the basic experience of the world of which science is the second-order expression." From Abram's perspective, it is only by returning to our senses, reawakening the forgotten intimacy and solidarity between the human animal and the animate Earth, that we have a chance of slowing and finally constraining the onrushing pursuit of knowledge and technological progress that we manifest at the expense of this breathing world.

Feeling: Arne Naess and the Deep Ecology Movement

Arne Naess, the distinguished Norwegian professor of philosophy, first saw the mountain Hallingskarvet in south-central Norway when he was seven years old. Even at such a young age he sensed that the mountain was a living being that emanated benevolence, magnificence and generosity. So great were these feelings that Naess vowed to live on his mountain as soon as he was old enough, for as long as he could. He held to his dream, and in his late twenties he built himself a cabin high up on the mountain, at the place called Tvergastein—the place of crossed stones.

During his long periods of living close to his beloved mountain, Naess gradually began to ask himself how the astonishing, and sometimes overpowering living qualities in the rock, wind and ice which encircled his eyrie high up above the tree line could help him to discover the *right way to live*. Naess's answer, which he calls 'deep ecology', aims to help individuals to explore the ethical implications of their sense of profound connection to nature, and to ground these ethical insights in practical action in the service of genuine ecological sustainability. The emphasis on action is what distinguishes deep ecology from other ecophilosophies, and is what makes deep ecology a *movement* as much as a philosophy. Perhaps the most fundamental insight of the movement is that all life has intrinsic value, irrespective of its value to humans.

For me, there are three radically interconnected senses of the word 'deep' in deep ecology, as shown in the diagram below (Figure 2). By working on these three aspects of deep ecology in oneself one can begin to develop what Naess calls one's own personal 'ecosophy', or ecological wisdom—a way of being in the world which minimises harm to nature whilst enhancing one's own feelings of awe, wonder and belonging.

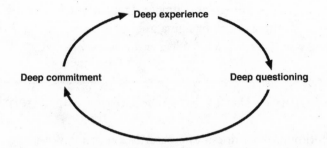

Figure 2: Three senses of 'deep' in the deep ecology approach.

First, there is *deep experience,* which is the sense of profound waking up into Gaia experienced by Aldo Leopold when he looked into the eyes of the dying wolf out in the mountains of New Mexico, and by David Abram in his Indonesian cave. Deep experience needn't be as dramatic as Leopold's or Abram's, nor is it felt only in wilderness areas, for there are many kinds of deep experience. Some people have them whilst tending small window boxes in the city; others experience a continuous, almost background sense of deep connectedness to nature. My own practices as a deep ecology educator have shown me that for most of us deep experience lies just below the surface of everyday awareness, and that a slight shift of context can easily make it visible. Deep experience is easily evoked, but its ethical implications are more difficult to assimilate.

This assimilation happens when one engages in *deep questioning* of both oneself and society. In questioning oneself, one asks whether one is living in a way which is consistent with the general flavour of one's deep experience by using the rational mind to tease out the web of connections between assumptions and actions at all levels of one's life in order to artic- ulate an ethical standpoint, which, although provisional and always under revision, can help to guide our lifestyle choices. In questioning society, one tries to understand its underlying assumptions from an ecological perspec- tive by looking at the collective psychological origins of the ecological crisis and of the related crises of peace and social justice. This deep ques- tioning of the fundamental assumptions of our culture contrasts markedly

with the mainstream shallow or *reform* approach, which tries to ensure
the continuance of business as usual by advocating the 'greening' of business and industry through a range of measures such as pollution prevention and the protection of biodiversity due to its monetary value as medicine or for its ability to regulate climate. Finally, one feels a sense of *deep
commitment* for the work of bringing about change in peaceful and democratic ways, which feeds back to deepen one's experience.

Thinking: James Lovelock and Gaia Theory

The perception of Gaia doesn't necessarily happen only through the gateways of intuition, sensing and feeling—thinking is an equally powerful
avenue. The most pre-eminent example of a person in whom Gaia has
manifested through an intellectual inspiration is of course the British
scientist and inventor James Lovelock, renowned for his theory of a self-
regulating Earth which he named after Gaia. Lovelock stumbled upon the
concept of a living Earth whilst working for NASA in the 1960s on the
problem of detecting life on Mars. Lovelock was well known as the
inventor of the electron capture detector (the ECD), an exquisitely sensitive instrument which provided the data for the shocking discovery, popularised by Rachel Carson in her book *Silent Spring*, that DDT and other
dangerous pesticides were widely distributed throughout the biosphere,
particularly in animals high up the food chain such as birds of prey.

NASA needed someone who would design a life-detecting instrument
that could be put on board a mission to Mars. It was clear to them that
Lovelock was well qualified for this task, but as it turned out, they had not
reckoned on his astonishing capacity for creative holistic thinking. The
scientists engaged in the life-detection project were trying to devise instruments which would sample the material on the Martian surface for Earth-
like organisms and their biochemical products, but Lovelock suspected
that such experiments would be of no use if life on Mars was biochemi-
cally and physically different to life on Earth, or if the lander happened to
sample a region of the planet where life just happened to be absent. As he
wondered whether an holistic approach might be more appropriate, a brilliant intuition suddenly bubbled up into Lovelock's mind: perhaps one
could detect life on Mars at the level of the whole planet by analysing its
atmosphere. After all, he reasoned, life radically alters the Earth's atmos-

Lovelock had received two startling insights. It was as if he had been an explorer in a far country who, reeling from his discovery of a wild sunlit river tumbling over a spectacular waterfall, had then stumbled, awestruck, through an opening in the trees to find yet another powerful cascade plunging over a precipice into the depths of the forest. The first insight was that life had regulated the composition of the atmosphere over geological time; the second, a logical extension of the first, was that life must also have regulated our planet's temperature.

Notice that the insights are described first of all in very general terms. Gaia (although he had not yet encountered that name) was a "revelation" and "awesome thought". This suggests that a deep experience might have been involved, akin to Leopold's conversion in the mountains of New Mexico, and indeed it was, but Lovelock's deep experience took the form of two powerful intuitions that he immediately translated into a scientific hypothesis. From then on, Lovelock's intuitive connection with Gaia was to grow gradually, like the slow growth of a copper sulphate crystal in super-saturated solution.

Excited by the astonishing idea of a living Earth, Lovelock tried to explain his idea to his NASA colleagues, but none of them really understood what he meant. He searched for a name for his insight, and whimsically played with the possibility of calling it the BUST hypothesis—the 'Biocybernetic Universal System Tendency'. Had he used this name, the idea of a self-regulating, living Earth might have easily caught hold in the scientific mainstream, but he chose to invoke the name given by the Greeks to their divinity of the Earth, which did not please the scientific establishment. There was, however, one eminent scientist who did respond with great enthusiasm when she first heard of Lovelock's idea—Lynn Margulis, the American evolutionist famous for providing conclusive proof that bacterial mergers about 2,000 million years ago gave rise to the complex cells we are familiar with today, such as those of mammals and plants. Margulis helped Lovelock flesh out his theory with many details about how microbes affect the atmosphere and other surface features of our planet, but the name 'Gaia' was given to Lovelock's notion of a self-regulating Earth by William Golding, an extraordinarily able man who was not only a Nobel laureate in literature, but also a physicist and a classical scholar.

One afternoon in the 1960s James Lovelock and William Golding walked together towards the post office in the village of Bowerchalke in

Wiltshire where they both lived. As they walked, Lovelock explained his vision of a self-regulating Earth to Golding, who was deeply impressed by the idea of our planet as a great living being. Feeling that this extraordinary notion had a great deal to do with the ancient Greek divinity of the Earth— with Gaia herself—Golding told Lovelock that this grand idea needed a suitably evocative name. Cautiously, Golding breathed the word "Gaia" into the enfolding air that swirled between the two friends like a secret listener at the doorways of consciousness. The Earth held her breath as the neurons in Lovelock's brain began to fire, for here at last was a possibility that her living presence could once again be recognised by the very culture that was at that precise moment laying waste to her great wild body. But Lovelock misunderstood Golding's suggestion, thinking that he must have been referring to the great 'gyres' that swirl over vast areas of the ocean and air. Prompted by a strange sense of urgency, Golding tried once more, making it clear that he was speaking of none other than the ancient and once revered Greek divinity of the Earth. Once more the word "Gaia" resonated in the air between the two friends, but this time Lovelock understood the name correctly, and a strange feeling came over him that he had at last found the name he had been looking for. But perhaps it wasn't just Lovelock who experienced delight, for it may well have been that in that very moment the multitudinous microbes, the great whales, the prolific rainforests, the wide blue oceans, the rocks and the air, and indeed the whole of life, all rejoiced in their knowing that this moment might at last signal the beginning of the end for our outdated, death-dealing objectifying view of the Earth.

Sensing the Round Earth

Lie down on your back on the ground in your Gaia place. Relax and take a few deep breaths. Now feel the weight of your body on the Earth as the force of gravity holds you down.

Experience gravity as the love that the Earth feels for the very matter that makes up your body, a love that holds you safe and prevents you from floating off into outer space.

Open your eyes and look out into the vast depths of the universe whilst you sense the great bulk of our mother planet at your back. Feel her clasping you to her huge body as she dangles you upside down over the vast cosmos that stretches out below you.

What does it feel like to be held upside down in this way—to feel the depths of space beyond you and the firm, almost glue-like support of the Earth behind you?

Now sense how the Earth curves away beneath your back in all directions. Feel her great continents, her mountain ranges, her oceans, her domains of ice and snow at the poles and her great cloaks of vegetation stretching out from where you are in the great round immensity of her unbelievably diverse body.

Sense her whirling air and her tumbling clouds spinning around her dappled surface.

Breathe in the living immensity of our animate Earth.

When you are ready, get up, breathe deeply, profoundly aware now of the living quality of our planet home.

Re-weaving the Covenant

Two sentences spoke out to me as I recently re-read Jacques Monod's famous book *Chance and Necessity*. "Animism," he says, "established a covenant between nature and man, a profound alliance outside which there seems to stretch only terrifying solitude. Must we break this tie because the postulate of objectivity requires it?" His answer was of course that we must, for he was a great believer in the truth of a soulless, mechanistic universe. According to Monod, it was the challenge of the West to break the ancient animistic covenant with the cosmos, and to learn to live with the consequences. But it is now becoming increasingly clear that the mechanistic view is literally killing the Earth as it was configured at the time of our birth as a species, and that in these desperate times our most urgent task is to find a way of re-weaving the ancient covenant with Gaia.

The moment when Lovelock accepted Gaia as the name for his insight might yet prove to be of huge significance for this enterprise and for the future life of our planet. For the first time in many thousands of years, the divine name which had once been used to revere the Earth as alive, sacred, and replete with meaning, character and soul has once again emerged full-blown and unashamedly into Western culture. Thanks to Lovelock and Golding, Gaia's long exile as a way of knowing the world is now over. At last Gaia has come home out of the outer darkness of our collective unconscious and into the light of modern awareness. Her name is now once again free to linger on the tongues and in the thoughts of countless numbers of people who seek to heal the blight inflicted on their lives by the objectivist fallacy of the West.

Gaia has had an increasingly profound impact both within the scientific mainstream and beyond, influencing a diverse range of fields such as philosophy, economics, health care and politics. Vaclav Havel, the Czech president, has pointed out that Gaia gives us all, whether atheist or believer, something much vaster than ourselves to be accountable to. On the political right, Margaret Thatcher made a speech in 1998 in which she said that before the century was out the environment would usurp the political agenda, and took action by establishing the Hadley Centre for Climate Research at the UK's Meteorological Office. She also tried to persuade her cabinet to read Lovelock's first book on Gaia, but unfortunately she failed to see how her vigorous promotion of unfettered economic growth helped to bring about the very environmental crisis that she claimed to so vehemently deplore.

As recent climatic disruption is showing us, we can affect Gaia in very serious ways, and she is indeed holding us to account by setting in train feedbacks that may well curtail and disrupt our activities long into the future. We are learning the painful way that we are embedded within a larger planetary entity that has personhood, agency and soul, a being that we must learn to respect if we wish to have any sort of comfortable tenure within her. Unlike human-made agreements such as business deals and international treaties such as the Kyoto protocol, our embeddedness in Gaia is non-negotiable. As Paul Hawken and his co-authors have noted, "Nature bats last, and owns the stadium."

The dawning realisation of the overarching importance of Gaia has spawned a range of new ways of relating to the Earth within Western culture. Many organisations covering a wide remit of green issues have

adopted her name, including Gaia House, the world-famous English meditation centre, and a host of foundations and charities around the world. I
see all of this as a good sign—it shows that Gaia as an archetypal idea is
capturing the imagination of people from many walks of life and from a
variety of psychological dispositions who feel disillusioned with the now
clearly outdated deterministic world-view which they see as needing to be
superseded by a more holistic approach. As David Abram has said, there
is no time to waste: the Gaian perspective must spread from sensing body
to sensing body like a contagion before it is too late to save what is left of
our endangered climate and beleaguered biodiversity.

Philosopher Mary Midgley has pointed out that in our own time, the
idea of Gaia is well placed to provide a powerful catalyst for the widespread adoption of holistic thinking in all spheres of life. The upshot of
her argument is that Gaia serves very well as a synonym for a mode of
thought capable of connecting seemingly disparate spheres such as science,
religion, politics, education, health care and crime prevention. In fact,
holistic thinking has been present in the science of ecology since its incep-
tion at the beginning of the last century. Ecologists have always realised
that the living beings they studied are interconnected in complex ways,
that the networks of relationships in ecological communities affect all the
players, that what happens to one species reverberates throughout the web
of life, affecting all.

Sadly, holistic thinking in the science of ecology has never really had
much of an impact on wider society. The appearance of Gaia theory has
the potential to change all that, perhaps because of the shocking discovery
that something as seemingly insignificant as the thin smear of life clinging
'passively' to our planet's surface is in fact no mere insubstantial
passenger, but is instead a major contributor to processes of titanic
proportions, such as the composition of the atmosphere, the global circulation of the oceans and the very weather patterns which we experience
every day as our most immediate connection with Gaia.

Perhaps this is because Gaia is not so small as to seem way beyond the
everyday world, and yet not large enough to boggle the imagination with
vast distances and far-flung cosmic events. A Gaian approach allows us to
develop a sense of how the local connects and contributes in multifarious,
surprising and significant ways to life at the global level. As we explore
some of her stories in this book, as we move through her various manifestations in our imagination, we will try to catch glimpses of the rich inter-

connections that are the very substance of her life, and will hopefully experience in ourselves a sense of wide identification with the whole of life in which we realise that we are truly inseparable from the living being of our planet.

Swimming Through the Atmosphere

Find a comfortable place to lie down on your back outside, preferably where there is native vegetation, and where you can look up through the canopy of branches and green leaves to the sky beyond.

Now invert your gaze, so that the ground behind you becomes the surface of a lake, and the air playing amongst the branches as you look down becomes water, deliciously translucent and fluid, from which you easily extract life-giving oxygen with every breath.

You are floating face-down on the surface of the lake. Trees have become water plants growing down into the silky water, and it is lovely to look down at them as you float on the surface.

Birds are fish darting through the water plants, or swimming through the deep clear waters beyond; falling leaves are gifts that make their way up from the depths to where you float on the lake's surface.

When you are ready, 'dive' down amongst the leaves and branches that grow into the clear depths of the lake like fronds of kelp. Plunge in amongst their leafy greenness and make contact with the water as it swirls around you.

You are swimming through your mother planet's atmosphere, made for you by a myriad of living beings, some living in the soil and in the ocean, others on the rocks and in the forests.

Return to the surface, and gaze once again into the crystalline water.

When you are ready, roll over and slowly stand up.

Back in the everyday world, breathe in the nourishing air, and, in turn, breathe out your own gaseous nourishment to the plants around you.

Chapter 3

From Gaia Hypothesis to Gaia Theory

Predecessors

James Lovelock was not the first scientist to speak of a living Earth. James Hutton (1726–1797), one of the fathers of modern geology, was responsible for discovering the cyclic nature of geological processes and the immense age of the earth, and is reputed to have thought of the Earth as a superorganism whose proper study was physiology. Lamarck (1744–1829) recognised that living beings were comprehensible only when seen as part of a larger whole. The Romantics, including Goethe, developed similar views, and it was Humboldt (1769–1859) who stressed the unity in nature and used the term 'Geognosy' for his holistic explorations of the Earth. Humboldt in particular saw the Earth as a great whole, and spoke of climate as a unifying global force, and of the co-evolution of life, climate and the Earth's crust.

In 1875 Eduard Suess published *The Face of the Earth* in which he imagined a space traveller discovering the surface of our planet. He spoke of the "solidarity of all life", and saw the Earth as a series of concentric envelopes—the lithosphere, hydrosphere, biosphere and atmosphere. Suess's ideas had little impact until the Russian scientist Vladimir Vernadsky (1863–1945) used the biosphere concept to develop a theory of the co-evolution of life and its non-living material environment. As a boy,

Vernadsky had been greatly influenced by his uncle, the philosopher Yevgraf Maximovitch Korolenko, to believe that the Earth was a living organism, and perhaps this led Vernadsky to see life as a "geological force" because of its ability to move matter in ways which geology alone could not. For Vernadsky, birds were nothing more than flying dispersers of phosphorus and other chemical elements, and human technology was an important part of nature because it increased the flow of matter and materials through the Earth's geology. Vernadsky thought that the smallest living beings—the bacteria—had the greatest influence on the chemistry and geology of our planet. These microbes could alter the Earth's crust, but, in the words of author Connie Barlow, "Only multicellular creatures are able to move matter in ways which wind and water refuse to carry it." Vernadsky didn't have any notion that the Earth might be self-regulating, but he did coin the term 'biogeochemistry', and his ideas gave rise to the Russian concept of the ecosystem, known as the biogeocoenose.

Lovelock, however, was the first to take the idea of a self-regulating earth far beyond the preliminary speculations of his predecessors, and was the first scientist to write books and scientific papers on the theme of Gaia. Lovelock's first mention of the Earth as a self-regulating system was in a scientific paper in 1968, before his crucial walk with William Golding. The first paper in which Gaia was explicitly used as the name for the self-regulating Earth was published in 1972. Further papers on Gaia were written with the help of Lynn Margulis in the early 1970s, and in these papers Lovelock and Margulis made bold statements such as ". . . life, or the biosphere, regulates or maintains the climate and the atmospheric composition at an optimum for itself." Lovelock called this initial formulation of his insight the *Gaia hypothesis*, because it was a tentative proposal which at that stage had very little data or theoretical underpinning to back it up. By suggesting that the biota were 'in charge' of the global environment, Lovelock and Margulis had turned the conventional scientific wisdom on its head, for as far as mainstream science was concerned, life was far from being in control—it was, in fact, nothing more than a second-class citizen on a ship captained entirely by the random forces of geology, physics and chemistry.

In this classical view, living beings had to either adapt to the environmental conditions set up for them by these great forces or be consigned to oblivion—they had no say whatsoever in where the ship was going. Later, when more data had become available, and after Lovelock had developed ways of modelling Gaia mathematically, the Gaia hypothesis evolved into

Gaia theory. The key insight of the theory is wonderfully holistic and non-hierarchical: it suggests that it is the *Gaian system as a whole* that does the regulating, that the sum of all the complex feedbacks between life, atmosphere, rocks and water give rise to Gaia, the evolving, self-regulating planetary entity that has maintained habitable conditions on the surface of our planet over vast stretches of geological time.

This is a radical departure from both the mainstream view which puts the non-biological processes in control of the Earth, and from the Gaia hypothesis, which had put life in charge. Gaia theory suggests that life and the non-living environment are *tightly coupled*, like partners in a good marriage. This means that what happens to one partner happens to the other, and implies that all the rocks on the Earth's surface, the atmosphere and the waters have all been deeply altered by life, and vice versa (Figure 3). The self-regulation arising from this tight coupling is an emergent property that could not have been predicted from knowledge of biology, geology, physics or chemistry as separate disciplines. Gaia evolves as an entirety and, like a beehive or a termite colony, is a superorganism, which for Lovelock is "an ensemble of living and non-living components which acts as a single self-regulating system". Thus the atmosphere is a much the product of life as is a cat's fur or the bark of a tree.

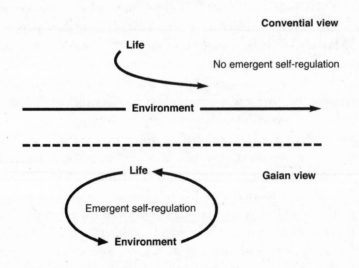

Figure 3: Non-Gaian and Gaian views of
the relationship between life and environment.

Not surprisingly, scientific orthodoxy reacted violently against Lovelock's hypothesis, declaring that it implied that some kind of mysterious global purpose maintained habitable conditions for the entire biosphere over vast periods of time. For mainstream science, to suggest that natural phenomena are purposeful in this way is to invoke the great heresy of *teleology*, the notion that nature is more than a great mechanism, that there is a mysterious purposefulness at work in the world, empowering it with the qualities of intentionality, mind and soul.

From a strictly deterministic perspective, this criticism was perfectly valid, and was welcomed by Lovelock because it spurred him on in his quest for a firm scientific foundation for Gaia. Many scientists, especially evolutionary biologists, railed against Gaia with a vehemence and disdain strongly suggestive of an irrational reaction against an idea with the potential to seriously threaten established mechanistic dogmas and disciplinary boundaries. Scientists objected to Gaia because they quite rightly perceived, perhaps unconsciously, that the idea implied a teleological view of the world, which, if accepted, would bring into question the fundamental belief that we can exploit this 'dead' old Earth of ours without restriction and with complete impunity. Of course, scientists could not be seen to object to Gaia on such inherently unscientific grounds; they needed to find good rational arguments for dismissing Gaia as bad science. The arguments were duly found and articulated.

There were three major criticisms levelled against Lovelock's earliest Gaian intuitions as he first expressed them in the Gaia hypothesis. Richard Dawkins, the famous Oxford evolutionary biologist, objected on the grounds that Gaia cannot be alive because there is no way that a planetary superorganism could come into being through the normal process of natural selection deemed by Darwin and Wallace to give rise to all the beautiful and complex forms of life we see around us. Darwin's theory suggests that for natural selection to work it is absolutely indispensable that parents have offspring which differ from each other, and that these variable offspring compete for scant resources so that only the best adapted survive long enough to pass on their genetic endowment to the next generation. Dawkins's point was that that Gaia could not have come about in this way since it is clearly absurd to think that natural selection could have operated amongst a variety of variable planets all springing forth from a single planetary parent.

A more important criticism came from the eminent evolutionary biologist W. Ford Doolittle, who, conceding for the sake of argument that planetary self-regulation does indeed exist, found it impossible to see how natural selection, operating at the level of selfish individuals concerned only with their own survival in local habitats and environments, could give rise to self-regulation at the level of the entire planet. For Doolittle, self-regulation, if it existed at all, was more a matter of good luck than an inevitable consequence of the relationships amongst living beings and their non-living surroundings.

Finally, there was the argument put forward by the climatologist Stephen Schneider, who pointed out that life and the non-living environment do in all probability mutually influence each other, but only in the form of a loose co-evolutionary dance. For Schneider, tight coupling between biota and environment does not exist, so there cannot be an inherent tendency for the emergence of planetary self-regulation.

These criticisms provided a Lovelock with a stimulating irritant that spurred him on to develop the strikingly original mathematical models of Gaia which we shall presently explore. But first let us look at the evidence for a self-regulating Earth.

The Evidence for Gaia

There is in fact good scientific evidence that our Earth does indeed have a remarkable ability for maintaining habitable conditions despite many external and internal forces that could easily have destroyed life during the course of her 4,600 million (4.6 billion) year existence. Look carefully at the graph in Figure 4, which shows that Gaia's average temperature has never been too hot or too cold for life, despite some warmer and some cooler periods. This relatively equable temperature history presents us with a considerable scientific puzzle because of the way in which our sun has evolved and changed during Gaia's lifetime.

Astrophysicists have established beyond reasonable doubt that our sun has gradually bathed Gaia with more and more energy over geological time, as more and more hydrogen has fused into helium in the sun's deep interior. The sun today is about 25% brighter than it was around 3,500 million years ago, when life first appeared on the planet, and yet Gaia's temperature has never been too hot or too cold for life. The data in Figure 4 come from hard science, and are surely excellent evidence for Gaia. To think that her relatively stable temperature could have been maintained over such immense periods purely by luck is tantamount to a highly unscientific belief in miracles. A more rational attitude would be to accept that the data are strongly suggestive of an innate tendency for planetary self-regulation, and to then ask how this might have arisen. The fact that Gaia has been able to maintain habitable conditions despite the external stress of an ever-brightening sun means that she is indeed in some sense alive, because self-regulation is a property of all living beings, or of their artefacts. You and I are no different from Gaia, in that we are able to keep our body temperatures within very narrow limits despite a wide range of external temperatures. We have an array of methods for doing this, which operate unconsciously. When it's hot outside we produce sweat, which a passing breeze evaporates, taking some of our excess heat with it. When it's cold, we redirect blood flow from the expendable extremities such as fingers and toes to vital internal organs such as liver and intestines, and we shiver to generate extra warmth.

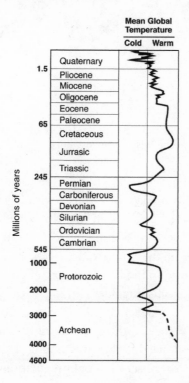

Figure 4: Gaia's temperature over geological time.
(re-drawn from K.C. Condie and R. E. Sloan, Origin and Evolution of Earth: Principles of Historical Geology, Prentice-Hall, 1998)

The message of a self-regulating Earth has also been found by scientists working in the Antarctic, who have extracted cores of ice from the frozen continent that contain air bubbles trapped by swirling snow up to 700,000 years ago (Figure 5, which shows a part of this record). By analysing the air held in these bubbles along the length of the core it has been possible to discover how carbon dioxide and temperature have varied together over this vast span of time.

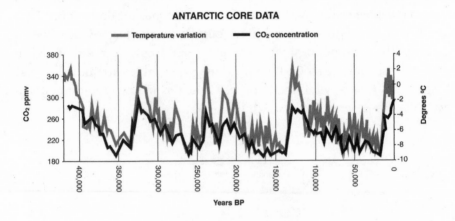

Figure 5: Gaia's temperature over the last 400,000 years, from the Vostok ice-core data.

Notice that there has been a regular rhythm in the traces of temperature and carbon dioxide, like a heartbeat or a pulse, with brief warm periods, such as the one in which our civilisation has flourished, every 100,000 years. A tiny change in the amount of sunlight reaching the Earth appears to set the rhythm because of the amplifying effects of complex internal feedbacks. Notice the close coupling between carbon dioxide and temperature, and also another astonishing fact: during each cold period, the amount of carbon dioxide has never dropped below 180 ppm (parts per million), and has never exceeded 300 ppm during warm periods. Even scientific sceptics now acknowledge that this is powerful evidence for a self-regulating Earth, for in the absence of tight regulation one would have expected the maxima and minima to vary much more erratically and unpredictably. Gaia has revealed herself through bubbles of ancient air.

Further powerful evidence for Gaia has come from studying the remains of life in rocks up to 550 million years old, the period of her history during which there has been multicellular life with body parts hard enough to be preserved as fossils. The results of this work are shown in Figure 6, where one can see five mass extinctions during which the diversity of life very rapidly declined, sometimes to alarmingly low levels as in the Permian extinction 250 million years ago, when about 95% of all fossilisable life forms vanished from the shallow oceans of the Earth.

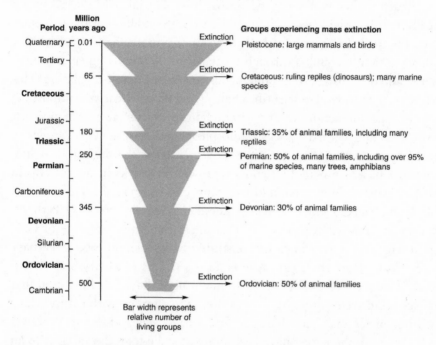

Figure 6: Diversity and extinction over the last 500 million years.
(redrawn from Richard B. Primack, Primer of Conservation Biology, Sinauer Associates Incorporated, 2004)

The last mass extinction happened about 65 million years ago, when a meteorite some 10km across struck the Earth in the Yucatan region on Mexico's Caribbean coast, followed by a huge flood basalt event in India. Opinion within the scientific community is divided over what caused the other four mass extinctions. Most experts favour meteorite impacts or changes in sea level and temperature wrought by the massive forces of plate tectonics; others believe that too much internal interconnection and complexity within ecosystems triggered these massive global losses of biodiversity. Notice, however, a striking fact: after each mass extinction it took 5–10 million years before the planet once again teemed with diverse life. This ability to recover from mass extinctions is strong evidence for Gaia.

Modelling Gaia: Daisyworld and Beyond

The evidence for Gaia did not convince the Doolittles of this world—as we noted earlier, they wanted to be shown how natural selection acting on locally selfish individuals could give rise to self-regulation at the planetary level. Lovelock knew that he had to produce a mathematical model of a biosphere tightly coupled through feedback to its non-living environment that would demonstrate the spontaneous emergence of self-regulation. This presented a considerable scientific challenge. It would clearly be impossible to model the interactions of up to 30 million species on our planet amongst themselves and with the rocks, atmosphere and oceans—the complexity was simply too vast to contemplate. Lovelock knew that he had to reduce Gaia to a basic template that could represent feedbacks in the real world without being so simple that it lost all connection with reality. He pondered this conundrum for about a year, without making any progress. Perhaps, he thought, the critics were right after all; perhaps the existence of Gaia could never be convincingly demonstrated with scientific rigour. Perhaps Gaia was destined to remain forever in the realms of poetry and philosophy. If so, the idea of a living Earth might never have filled the need that our science-driven culture has created for an integration with nature which satisfies our reason, as much as our intuition, sensing and feeling. As Lovelock cogitated, there may have been more at stake than the survival of an abstruse and remote scientific concept of interest only to experts in their ivory towers. Perhaps it is not too melodramatic to propose that, as Lovelock grappled with this problem, the future of our own culture, and of Gaia as we know her, hung in the balance.

Feedback

One of the key concepts for developing a rational understanding of Gaia is the notion of feedback, which was formally developed as the science of cybernetics by Norbert Weiner and others in the 1940s and 50s, but which in fact goes back much further back to inventions such as James Watt's steam governor, the regulators used to control the speed of windmills and the float valves used to regulate the speed of Greek and Roman water clocks. The very word 'feedback' is redolent of the notion that nature is nothing more than a deterministic set of complex interacting parts, so I prefer to breathe a sense of animism into the notion by thinking of feedback loops as *circles of participation*—as manifestations of the ways in which the deep, awesome sentience of nature organises itself into meaningful relationships that bring either constancy or change. But to abandon the terminology of cybernetics in favour of more participatory language could be difficult for readers steeped in the language of science. Thus, I continue to use the term 'feedback loops', but will ask you to remember that they are merely reason's way of depicting the mysterious, animate *tai chi* of life, the incessant dance of existence.

A system with feedback is one in which a change in one component of the system propagates around a loop of interrelated components until, eventually the original component experiences a change. Feedback loops can be either negative or positive. In a negative feedback, the initial change is counteracted, whilst in a positive feedback the initial change is amplified.

First, let's look at a system with simple negative feedback (Figure 7):

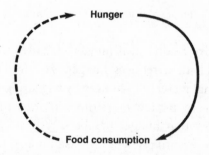

Figure 7*: A system with simple negative feedback.

Here we have only two 'parts' or components that interact, namely hunger and the consumption of food. Systems theorists have worked out a very simple but highly effective notation for describing feedback that we shall be using throughout this book in our explorations of Gaia. Notice that

---

* Figures 7, 9 & 12 are re-drawn from *An Introduction to Systems Thinking*, Stella Research Software, 1997.

hunger and food consumption are connected with arrows, which denote *couplings*. These are of two types: solid and dashed. A solid arrow denotes a *direct coupling*, in which an increase in the component at the tail of the arrow causes an increase in the component at the tip of the arrow, and vice versa. Conversely, a dashed arrow denotes an *inverse coupling*, in which an increase in the component at the tail of the arrow causes a *decrease* in the component at the tip, and vice versa (Figure 8).

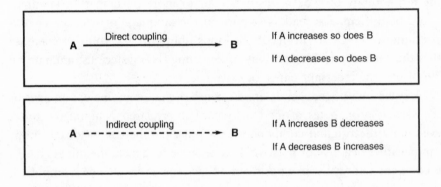

Figure 8. Two kinds of couplings.

In our simple negative feedback, if hunger increases then food consumption increases, which then reduces hunger. The feedback has therefore counteracted an initial increase in hunger by triggering food consumption. Let's flow around the loop a few more times in order to explore its longer-term behaviour. Now that hunger has been reduced, food consumption will go down, which eventually will increase hunger, bringing us back to the situation at the beginning. No matter how long we flow around the loop, hunger and food consumption will oscillate around average values, and will never increase or decrease without limit. This loop, like all negative feedbacks, is self-regulatory.

The other basic cybernetic relationship is positive feedback (Figure 9). Here is a diagram of a simple system of this kind.

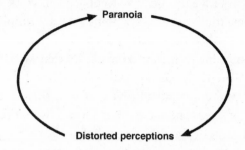

Figure 9: A system with simple positive feedback.

In this example, there are only solid arrows. This means that if my perceptions become distorted, my paranoia grows, which then feeds back to increase my distorted perceptions, and so on. The result is that both paranoia and distorted perceptions increase without limit. This is a classical 'vicious' cycle, since the system has no tendency for self-regulation. Imagine that my distorted perceptions cause me such distress that I decide to talk about my situation with a close friend, or I might choose to see a therapist. Either way, if the help is successful in reducing my distorted perceptions, my paranoia will be reduced, and this will have the effect of reducing my distorted perceptions. Now the change is in the opposite direction—instead of an unlimited increase, we now have a potentially unlimited decrease—a 'virtuous cycle' that leads me into increasing levels of sanity and well-being. Notice that there is no emergent self-regulation in positive feedback; there is only constant change, either towards more and more, or towards less and less.

One technically inappropriate—but nonetheless effective—way of remembering how negative and positive feedbacks behave is to imagine that a system's sign reflects its *attitude* to change. A system in negative feedback, when presented with change, is 'negative' to it, preferring instead to stay where it is. A system in positive feedback, on the other hand, loves change, and is hugely 'positive' towards it. There is a simple rule of thumb that we will find very useful for deciding if a feedback loop is in negative or positive feedback: simply count the number of inverse couplings, that is, dashed arrows. If there is an odd number, the feedback is negative, if there is an even number, or none at all, then the feedback is positive.

In fact, neither negative nor positive feedback can work without sensors that detect tiny deviations from a set point that are amplified before the signal is fed back into the original component. In Gaia, much of the amplification comes about due to the amazing ability that all living beings have for exponential growth. A classic example comes from the world of bacteria, in which the unrestrained divisions of a small initial population would in a matter of days generate so many new cells that their mass would equal that of the Earth. In Gaia the exquisitely delicate receptivity of living beings to their surroundings acts as an environmental sensor for the planet as a whole. Systems theorists usually denote an amplifier by means of an inverted triangle, where the ingoing arrow denotes the input signal, and the outgoing arrow the amplified output (Figure 10):

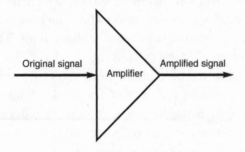

Figure 10: A simple amplifier.

Using this notation, a Gaian feedback would be drawn like this (Figure 11):

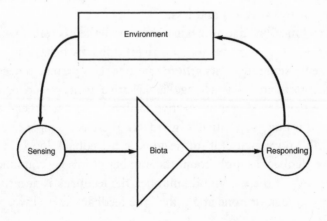

Figure 11: Amplification with feedback.

The physics and chemistry of the environment influence living beings (the downward arrow), which use their delicate sensory abilities to decide to what extent the outside world has deviated from the optimum for their own flourishing. They respond by amplifying the difference (the triangle) through active feedback, thereby shifting environment to a state more conducive for their own growth and well-being (the upward arrow). Trying to draw complex feedback diagrams with this notation would makes things very cumbersome, so I ask you to remember that sensors and amplifiers like these are 'hidden' inside of many of the components of the Gaian feedback diagrams that we will be working with. Life-mediated amplification makes the feedbacks in Gaian models far more responsive and unpredictable than those of conventional models of the Earth, which treat life as a simple 'black box' in which little sensing and amplification takes place.

It is possible to generate very complex systems which include both positive and negative feedbacks, and this is what scientists charged with making models of future climate try to do on their supercomputers. Let's look at a more complex model now that links both positive and negative feedbacks as an illustration (Figure 12).

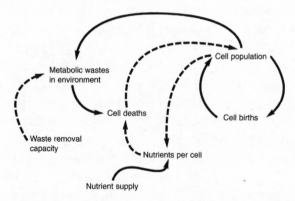

Figure 12: A tumour as a complex system with both positive and negative feedbacks.

This example makes the point that complex systems consist of veritable spaghetti junctions of relationships between many components. What makes a tumour so potentially deadly is the positive feedback, which can lead to runaway exponential growth. But there are two negative feedbacks which could restrain or even remove the tumour if they are strong enough. If the tumour receives a fixed supply of nutrients from the surrounding tissues, more tumour cells will die of starvation whenever the tumour attempts to grow, thereby reducing the tumour cell population. Likewise, if there is a limited but constant capacity to remove toxins made by the tumour, then as this grows more of its own cells will die of self-poisoning, which will reduce the size of the tumour. How the system behaves will depend entirely on the quantitative relationships between these three feedbacks.

Wherever we look in the biological world we find astonishingly complex feedbacks of this kind, whether it is within the physiology of individual organisms, in the ecological interactions within ecosystems, or indeed amongst the interactions between rocks, atmosphere, oceans and living beings that constitute Gaia. Such complex systems can behave in unpredictable ways. The precise behaviour will depend not only on which relationships are present, but also on their relative strengths. Invariably, non-linear relationships will be present, and if so the system will exhibit a range of behaviours, from predictable to chaotic, but with sufficient complexity even a system with linear relationships can give rise to all sorts of behaviours. Simply put, in a linear relationship, a component's response varies in direct proportion to a change it experiences, and in a non-linear system it doesn't. A good example of non-linearity is the stock market, where a slight change in consumer confidence can ripple through the system very quickly to bring about rapid and unexpected change. It is also possible for tipping points to exist, in which a small disturbance triggers a sudden and unexpected change. Take a pencil and line it up parallel to the edge of a table, not too far from the edge. Now give the pencil a slight push towards the edge and nothing dramatic happens—the pencil has moved a little, and just as you expected, it is still on the table. Now give it another small push, and another. Again you observe a predictable response. Eventually, of course, another slight push equal to all the previous ones takes the pencil through a rapid tipping point as it falls over the edge and into a new 'stable state' on the floor. Non-linear systems are riddled with tipping points, but often a system is so complex that it is impossible to know exactly when these will be encountered.

Daisyworld

We are now well prepared to return to Lovelock and his struggle to create a convincing cybernetic model of a self-regulating planet. On Christmas Day 1981, as Lovelock pondered his critics' points without conclusion, a strange coincidence happened which gave him the lead he was looking for.

On his desk was an open copy of the journal *Nature*, one of science's most prestigious periodicals. Lovelock was about to put it back but just happened to glance at the article spread out before his gaze. To his amazement, it described some straightforward equations for the growth in grasslands and lawns of the plantain, a common English wild plant. As he contemplated the text a flash of inspiration leapt into his mind. Into his consciousness came a visualisation of a planet with only two species—light- and dark-petalled daisies—representing the vast biodiversity of our planet. The model daisies would grow according to the equations in the *Nature* paper and would compete for space on the planet's surface, thereby satisfying the evolutionary biologists' requirement that competition be part of Gaia. But the innovative move that Lovelock made in his model was to wire in an explicit mathematical feedback between life—the daisies—and their non-living environment—the surface temperature of the planet.

Here we need a digression to explore the concept of *albedo* (from the Latin *alba*, meaning white), which, on a scale ranging from zero to one, scientists use as a measure of how much solar energy is reflected by a surface. When exposed to the sun, a high albedo (light) surface reflects most of the solar energy striking it without any transfer of energy taking place, whilst a low albedo (dark) surface heats up because it absorbs solar energy. The heat then radiates out to the surrounding air, warming the immediate vicinity, and, ultimately, the entire planet.

In Daisyworld, dark daisies warm themselves and the planet by absorbing energy from the sun which they radiate into their surroundings as heat, whilst white daisies cool themselves and their world by reflecting solar energy back to space. Lovelock constructed these couplings between life and the non-living environment using well-known equations from thermodynamics, the branch of physics that deals with energy flows. Also included in the calculation of the planet's surface temperature was the albedo of the bare ground, which was set, realistically, at a uniform grey. He then modelled the tight coupling between the planet's non-living environment (the surface temperature) and the growth of the daisy species by giving both the same realistic, bell shaped response to the planet's surface temperature. Both grew best, as real plants do, at 22.5^0C, and not at all below 4^0C or above 40^0C. Finally, as for the real Earth, Lovelock modelled the sun to gradually increase its output of energy as time went by. Figure 13 gives the flavour of how Dasiyworld works by focusing, for the sake of simplicity, on how only white daisies regulate the planet's surface temperature.

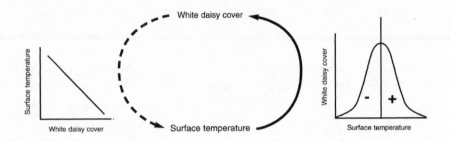

Figure 13: Feedbacks between white daisies and the surface temperature of their planet. On the left is the effect of white daisy cover on surface temperature. This is an inverse coupling: for a constant solar luminosity, as white daisy cover increases, surface temperature decreases. On the right is the bell-shaped effect of surface temperature on white daisy cover: to the left of the optimum growth temperature (the negative feedback regime) the coupling is direct; to the right of it (the positive feedback regime) the coupling is inverse. The brightness of the sun (not shown) in part determines which feedback regime applies.

What Lovelock had done was to create a model with six tried and tested equations from physics and biology to represent a simple but nevertheless realistic *feedback* between life and its environment. In doing so, he had created a new way of modelling the Earth as an integrated system of life and climate. Daisyworld was a bold and novel move—no one had ever attempted this, despite the injunction in 1925 by Alfred Lotka, the founder of theoretical population ecology, that species should not be modelled without reference to their non-living environment. As he prepared to run the model on his computer for the first time, Lovelock had only a faint inkling of what might happen. Daisyworld was a shot in the dark—had it failed, Gaia might well have languished in the obscurity accorded to quaint but now irrelevant scientific concepts, such as the theory of phlogiston.

As it happened, Daisyworld produced a startling result (Figure 14). As the young, cool sun shone down on Daisyworld, dark and light daisies began to grow as soon as the bare ground became warm enough to favour germination. Under the cool sun, dark daisies warmed themselves closer to the optimum growth temperature, and so they spread, out-competing the light daisies that cooled themselves into oblivion. Dark daisies grew explosively into the barren expanses of the planet through positive feed-back, darkening its entire surface and warming the world exponentially fast as they spread. As time went by, Daisyworld entered a long, stable period of negative feedback. Light daisies thrived increasingly well, because under a brightening sun they gained a slight advantage over the dark daisies by being able to cool themselves closer to the optimum growth temperature. When the sun reached its middle age, there were equal numbers of dark and light daisies. As the sun's brightness increased even further, dark daisies slowly vanished as they warmed themselves far above the temperature for optimum growth, leaving room for an expanding coverage of the better-adapted light daisies.

What amazed and delighted Lovelock most of all was the thin red line on his computer screen that traced the overall temperature of the planet as the sun increased its output of energy. The line revealed two startling emergent properties: first, the overall temperature of the model planet had remained remarkably constant over a vast period of time despite the shifting populations of daisies and an ever-brightening sun, and second, the temperature had settled on a value just below the optimum for daisy growth.

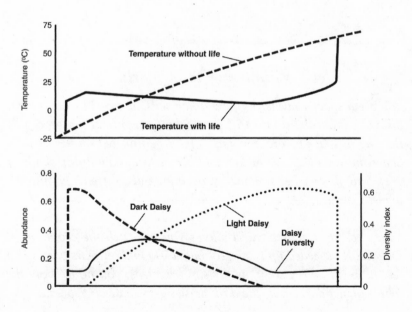

Figure 14: The original Daisyworld.

But eventually, under an old and exceedingly bright sun, the whole world was totally covered with light daisies, at which point it could not become any lighter. The sun, continuing its relentlessly increasing output of energy, finally warmed the now totally white Daisyworld above the optimum temperature for daisy growth. Lovelock watched, fascinated as the white daisies, now in positive feedback, began to die off exponentially fast, leaving more and more bare ground exposed to the sun. The planet's temperature rapidly soared with the demise of the white daisies, until it reached the high, lifeless temperature determined solely by the interaction between the bare ground and the energy from the sun. Daisyworld had died.

Lovelock had discovered that the feedbacks between the daisies and their non-living environment had allowed Daisyworld to regulate its own surface temperature close to the optimum over a vast period of time without the need for a mysterious disembodied teleological entity managing the planet behind the scenes. Lovelock had demonstrated that emergent self-regulation at the level of an entire planet was a distinct possibility in the real world.

Visualising Daisyworld

You are a space traveller looking down on Daisyworld from a great distance. From the comfort of your cabin you gaze at her gracefully floating in space before any daisies have germinated on her surface. She is the size of a large melon, and her surface is a uniform shade of grey. You also see her sun nearby, beaming out solar energy onto the planet's surface.

Your capacity for time travel allows you to watch the evolution of Daisyworld speeded up vastly, so that 3,500 million years passes in a few moments. Very soon, as the sun brightens, you notice how the whole planet darkens as the dark daisies germinate and spread.

Then, as the sun warms further, you observe the planet lightening as white daisies germinate in amongst their dark competitors. You watch the planet gradually becoming lighter and lighter in step with the brightening sun, until eventually the entire world is pure white, shining like a pearl under the intense solar energy.

Then, as the sun brightens just a little more, you watch the white daisies rapidly disappear, driven to extinction by the intense energy of the sun.

Daisyworld is dead, and has returned to a uniform shade of grey.

Daisyworld has become the mathematical basis for making more sophisticated models of Gaia, such as the new generation of long-term climate prediction models being made at the Hadley Centre and elsewhere. Like Daisyworld, these models have four basic properties: they include organisms that exploit opportunities for exponential growth; there is natural selection to ensure that only the most successful organisms pass on their traits to the next generation; the physical environment imposes constraints on the growth of the organisms; and, lastly, the organisms in turn profoundly affect their physical environment. These four conditions are Lovelock's "recipe for a Gaian system".

Daisyworld displays many interesting features. Its emergent regulation of surface temperature is remarkably resistant to perturbation of various sorts, including plagues that kill off a large proportion of the daisies, and sudden temporary reductions in the intensity of the sun. Another interesting feature of the model that may have disturbing implications for the real Gaia is what happens at the moment when Daisyworld dies of overheating. Just before life disappears, the light daisies cope with small increments of solar energy by increasing their cover of what little bare soil remains. But under a very bright sun, with no more bare soil available, a small increase in solar energy extinguishes life with sudden rapidity. A similar event takes place in more complex versions of the model in which daisies occupy a three-dimensional, spherical planet. It seems that small disturbances at critical moments in the life of a self-regulating system can trigger unexpected and sudden shifts into totally new system configurations. Could it be that just an extra increment of pollution or habitat destruction could trigger an equally dramatic shift towards a new and potentially inhospitable climatic regime on our real Earth?

Lovelock's next move with Daisyworld was to run the model with many daisy species, each with a slightly different albedo on a scale of almost pure black to pure white. Once again, the model regulated temperature beautifully. When he added rabbits that ate daisies and foxes that ate rabbits he expected to see wild fluctuations of temperature, but to his immense surprise he found that this more ecologically complex world was remarkably stable. My own work on Daisyworld, conducted with Lovelock as a guide and mentor, involved using the model to explore the vexed question in ecology of whether more complex ecological communities—those with more species and more interconnections between species—are better able to recover from disturbances than simple communities. This area of ecology, known as the 'complexity-stability debate', has been of central concern to ecologists since the 1950s, when the pioneer ecologists Charles Elton and Robert MacArthur suggested that more complex communities should be more stable. Their reasoning was based on various lines of evidence, including the fact that islands with fewer species seemed to be more vulnerable to invading species than species-rich continental areas, that crop monocultures are more vulnerable to pest outbreaks, and the apparent lack of insect outbreaks in species-rich tropical communities compared to boreal and temperate communities.

This was the accepted wisdom for many years, until the Australian theoretical ecologist Robert May published the first mathematical analysis of this problem. May turned Elton and MacArthur's insights, and the whole of scientific ecology, inside out by showing that more complex communities were far more likely to collapse than simple ones. Further modelling by May's successors did little to overturn these counter-intuitive results, and the controversy raged for about thirty years. Early on in my work on Daisyworld I realised that it provided an interesting template for exploring this issue for the simple reason that no one in the whole 60-year history of theoretical population ecology had attempted to address the complexity-stability debate by modelling the feedbacks between life and its non-living environment along the lines that Lovelock had developed.

With Lovelock's help, I created a new version of Daisyworld populated by 23 daisy species along a gradually increasing gradient of albedo, from light to dark, and then introduced three herbivore species which, unlike Lovelock's earlier daisy-eaters, behaved more like real predators by largely ignoring rare prey and feeding mostly on the commonest daisy species. I manipulated the complexity of my food webs by changing the number of daisy species eaten by each herbivore. In the simplest food webs, each daisy was assigned only one randomly selected herbivore out of the three as its predator, whilst in the most complex food web each daisy was eaten by all three herbivores. The stage was set for a Gaian exploration of the complexity-stability debate.

One dull winter's day whilst working at my computer at Schumacher College in Devon, I programmed the most complex food web, and prevented the herbivores from eating any daisies until almost a quarter of the way through the run as a control. With no predation, the model rapidly settled into its usual stability with constant populations of two dominant daisy species and constant global temperature. Then I set the herbivores loose, and running amok as much as equations can, they followed their mathematical destiny and decimated the two abundant daisy species, thereby opening up space for the many other species previously present only as seeds in the rich soil. The herbivores had abolished a floral tyranny of the minority, replacing it with a massive and permanent increase in daisy diversity as species of all albedos flourished, each at a low but constant abundance, with none overly dominant. The astonishing thing was that all of this had virtually no effect on the planet's overall temperature that displayed itself as a stable flat line on my computer screen, seemingly oblivious to the convulsions and upheavals going on amongst the populations of daisies and herbivores.

Then I programmed a low-complexity food web, in which each daisy species was eaten only by a single randomly selected herbivore. As before, I began by holding the herbivores in check and watched as two daisy species again reached high constant abundances and as the global temperature reached a constant level. Then the herbivores were set free. To my utter amazement, the stability vanished, replaced by wild rhythmical oscillations of temperature, daisies and herbivores. These effects were repeated for most solar luminosities and random connectance patterns. Tightly coupled feedbacks between complex biological communities and the non-living environment had stabilised both global climate and population dynamics in my model Gaian ecosystem (Figure 15). In all of these experiments, adding a herbivore-eating carnivore knocked back the dominant daisies and restored stability. Modelling such as this, combined with extensive evidence from the field, supports Elton and MacArthur's early intuitions that more complex communities are indeed, in general, more stable.

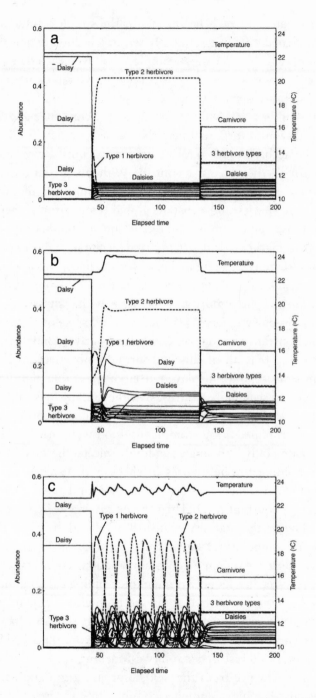

Figure 15: Ecological Daisyworld with a: high complexity,
b: intermediate complexity, c: low complexity.

Other theoretical ecologists, intrigued by the growing acceptance of Gaian thinking in science, and particularly in climatology, have begun to make their own models in which life and environment interact. One of the most eminent was none other than the late William Hamilton, who was responsible for generating the theoretical insights later popularised by his colleague Richard Dawkins under the rubric of the 'selfish gene'. Hamilton, in collaboration with ecologist Peter Henderson, developed a model called 'Damworld', set in a steep valley through which flows a tumultuous river with a narrow outlet in which nutritious algae grow. A second species, a dam builder, roots itself in the sediments around the outlet and feeds on the algae. As the dam builders multiply, the dam grows and so does the lake behind it, creating a great variety of new habitats for all sorts of creatures that weave themselves into a complex emergent ecological community as they are added one at a time. Some of these new species are dam busters that chew up the dam to extract the nutritious organic binding material that holds it together. Others eat the dam busters, and are eaten in turn by their own predators. Thus there is potential for the dam to rise and fall repeatedly over time, and it often does, but the whole system is far more likely to end up in a stable state with a high dam and a large lake when the links between the organisms and their non-living environment (the dam) become tighter and more pervasive. As in Daisyworld, tight coupling between living beings and their material environment leads to stability.

Chapter 4

Life and the Elements

Atoms as Beings

In order to deeply understand the life of our planet we need to explore the cycles of the chemical elements, for without their coming to life in organisms there would be no Gaia to speak of. So what exactly are the chemical elements? Democritus was right—the material world is indeed made of atoms, but, as we have seen, atoms are not dead, mechanical entities; they are participatory beings with characters akin to our own, even though those of atoms are far more consistent than human nature, which is malleable, often unpredictable, and very much dependent on circumstance. Place the same person in the same environment today and tomorrow, and on one day they may feel good, on the other they may feel sad. According to mainstream science, atomic personalities are far more consistent—an oxygen atom will *always* behave in the same way whenever it encounters the same chemical and physical environment, no matter where in the universe the encounter happens. Every person (and indeed, every mammal, bird, reptile, and possibly insect) has a unique character— no two hummingbirds will behave in exactly the same way, but every oxygen atom is considered to have exactly the same response, the same character everywhere throughout the universe.

This does not mean that atoms, or the protons, neutrons and electrons that constitute them, are no more than totally isolated, independent, self-existing entities interacting like billiard balls in totally predictable ways. Modern physics shows us that nothing in the universe exists in splendid isolation—everything depends for its very existence on its relationships with everything else. When a hydrogen atom bonds with an oxygen atom, aspects of the personalities of hydrogen and oxygen are brought out in the relationship which are not present in oxygen and hydrogen alone. This particular combination, known as the hydroxyl ion, is gaseous at ambient temperature, and is extremely good at scrubbing pollutants out of the atmosphere. However, when *two* hydrogen atoms bond with an oxygen atom, a totally different set of qualities spring forth in a new emergent domain with all the extraordinary and life-giving properties of water with which we are all so familiar. So atoms have freedom, but far less freedom than us. If we think of matter in this way, we can draw two important conclusions. Firstly, it no longer seems to be such a stretch to imagine that the many little 'freedoms' in matter can produce the high level of freedom that is so much a part of human consciousness. In the words of philosopher J. McDaniel, "Our own subjective experiences are highly developed forms of what there was in the beginning in sub-micro-scopic matter", and "'Matter' and 'mind' are simply names for different types of actual occasions of experience." In the words of philosopher Christian De Quincey, "Matter tingles with experience" and "Matter *feels* to its deepest roots." Secondly, we can no longer treat matter with disrespect, because it is, after all, *sentient* in some sense by virtue of having a creative agency and capacity for experience that demands our ethical consideration. We realise the profound wisdom in the etymological root of the word 'matter', which comes form the Latin for 'mother' (*mater*), and 'matrix', or womb.

In Western philosophy, the idea that matter is sentient is sometimes referred to as 'panpsychism'. The differences between panpsychism and the new animism that we considered earlier are not clearly defined, but one could tentatively say that panpsychism is more concerned with exploring the cognitive or rational implications of the insight that the world is, in Thomas Berry's words a "communion of subjects", whereas the new animism places more emphasis on working out how we should behave in relation to these subjects by rooting ourselves more explicitly in the perceptions of indigenous peoples.

If this approach is correct, then, in contrast to the mainstream view, we can conceive of matter as being inherently creative. Matter falls into certain patterns of relationship improvisationally, much as an artist explores new domains of being and interaction. For animists, matter and psyche are indissoluble, for the psyche of the world resides nowhere else but in matter itself. Thus the great archetypes of Gaia and *anima mundi* that figure so importantly in the human soul could well be prefigured in some mysterious way not in some abstract realm far from this world, but in the very molecules and atoms that constitute our palpable, sensing bodies. Perhaps psyche becomes visible when the relationships amongst a community of interacting agents are powerful and complex enough to call it forth from within the very matrix of materiality. If it is true that psyche is indeed revealed in the very thick of relationship, then Gaia may well be a domain in which the presence of living beings so quickens and intensifies the planet-wide interactions amongst atoms, rocks, atmosphere and water that the Earth literally awakens and begins to experience herself as alive and sentient.

Thus, panpsychism and the new animism teach us that chemistry need no longer be thought of in merely mechanical ways, as if 'chemicals' are nothing more than dead, static cogs. Chemical properties are sublimely fluid—they are the ways in which different aspects of the inner natures of the elements reveal themselves in different contexts and circumstances, just as our own behaviour is dependent to some extent on the social setting in which we find ourselves. But we must be careful when speaking of 'atoms'. The danger, alluded to above, is that we begin to think of them as things that exist in isolation from everything else, even whilst they interact with each other.

There are many different kinds of atoms, each with different chemical personalities. These are called the *elements,* because they were once thought to be the building blocks of everything around us, including our own bodies, and of course, the Earth. In the last century it was discovered that atoms are in fact not fundamental after all, but are themselves divisible into three 'particles'—the electron, the proton and the neutron. The physicist Neils Bohr was responsible for providing us with a now outdated but nevertheless useful model of how these particles are arranged within atoms and how they interact with each other. Atoms, he said, are like miniature solar systems. In the centre, in the place of the sun, is the nucleus, which is composed of positively charged protons and charge-less neutrons. Around the nucleus are the negatively charged electrons, orbiting around it as planets do around the sun. Protons and neutrons are heavy, and constitute most of the mass of an atom, whilst electrons are almost without mass at all, but all atoms are 99.99% empty space. All the particles wink in and out of existence according to some quantum physicists, entering what cosmologist Brian Swimme calls the 'All Nourishing Abyss'—the quantum vacuum or zero point energy field, when they briefly pass out of existence.

Despite their immense difference in mass, an electron's negative charge exactly cancels out the positive charge on a proton. Bohr pointed out one further astonishing fact—that the numbers of protons and electrons in pure elemental atoms are always equal, so that the overall charge in such atoms is zero. Science seems to gloss over an extraordinary fact—that 'positive' and 'negative' exist at all. What a mystery it is that like charges repel, and unlike attract. The fact that these two polarities must always try to be close to one another, to 'cancel' each other out, or to 'complement' each other— we know not which—gives us all of biology and chemistry, and quite a lot of physics. The French say that '*le contraire se touche*', and Aristotle said that two mysterious forces run the universe—attraction and repulsion. He saw in this the action of *anima mundi*, the ineluctable interiority that lies at the heart of matter-energy. We really can't do much better than Aristotle when trying to get to the bottom of this, as it leads us nowhere to reduce positive and negative charges to something more 'fundamental'; if one attempts this kind of futile reductionism, one ends up with a highly unsatisfactory situation of infinite regress. Attraction and repulsion have something to do with the intelligence, with the 'soul' of the universe itself—they are the manifestation at the level of matter/energy of the participatory nature of electrons and protons, perhaps no different in principle to the attractions and repulsions that we humans feel towards each other.

Thus, atoms, like humans, are constantly trying to find fulfilment. We find it in all sorts of complex and various ways. For me, there is nothing better than walking out in a wild, free, wildlife-rich, sun-soaked landscape, far from roads and the hubbub of modern culture. For others, it might be standing for hours in a crowded stadium in the English winter screaming encouragement to eleven men in shorts trying to kick a ball into a net more frequently than an opposing gaggle of eleven men. For an atom, things seem to be much simpler and more consistent—they *all* find fulfilment by arranging things so that they have the right number of electrons orbiting their nucleus.

Bohr suggested that an atom's electrons are arranged in a set of concentric orbits. The innermost orbit can hold a maximum of two electrons, the next one out, eight, the one beyond that another eight, the next 18, then 32, and so on for the outer orbits which needn't concern us here. Electrons in orbits closer to the nucleus have less energy than those further out. The whole of chemistry, and the whole of life, depend on the simple fact that every atom is utterly compelled to do whatever it can to end up with a full outer orbit. Atoms aren't satisfied until they achieve this, and of course atoms can't do this alone; they have to interact with each other to share or exchange outer electrons, and in so doing they create the bewildering variety of *molecules,* or communities of atoms, that we see in the worlds of chemistry, physics and biology. Each molecule is in an emergent domain with properties not reducible to those of its constituent atoms. Water is a good example. Two parts hydrogen and one part oxygen, its melting point, fluid dynamics, expansion on freezing and so on cannot be fully predicted from a knowledge of oxygen and hydrogen separately.

Gaia's Elements

The most important elements for life and Gaia are just six: carbon, hydrogen, nitrogen, oxygen, phosphorus and sulphur, remembered in the trade by the acronym CHNOPS. Let's look at how Bohr's model helps us to understand the personalities of these six chemical beings. Carbon has a full inner orbit with two electrons, but the next and outermost orbit is incomplete, with only four (Figure 16). Thus to find completion, carbon needs another four electrons, and finds them by sharing those in the outer orbits of other atoms, especially other carbons, forming what chemists call *covalent* bonds. When this happens, both atoms at last achieve a satisfactory resting state. The fact that carbon has a need for four electrons makes it a highly cooperative and intensely social chemical being. It is the solid, reliable Swede of the chemical world, loving nothing better than to share each electron with a fellow carbon atom, which means that it can link up with four neighbouring carbons to make large chains, rings and chains of rings in which multitudes of carbon atoms and associated oxygen, nitrogen, phosphorus and other atoms find collective fulfilment in the huge, often complexly convoluted molecules of life such as sugars, proteins and DNA. These linkages amongst carbon atoms are the basis of life as we know it; without them Gaia could not exist, and our planet would be as devoid of living beings as is our nearest neighbour, the moon.

The essence of carbon is centrality. It occupies a pivotal place in the community of elements by virtue of its half-filled outer electron shell (Figure 16), and it also occupies a central place in the workings of Gaia: atmospheric carbon dioxide and methane help to set the global temperature; dissolved carbon compounds regulate the acidity of the oceans; and, as we shall see, the burial of organic carbon helps to regulate the oxygen content of the air. The key point about global temperature is this: any gas molecule in the air that contains two or more atoms delays the escape to space of heat given off by the Earth after it has been warmed by the light of the sun. Carbon is present in at least two such *greenhouse gases*: carbon dioxide (CO_2, Figure 17) and methane (CH_4). Water vapour (H_2O) is another potent greenhouse gas. The addition of these and other greenhouse gases to the air raises global temperature; their removal decreases it.

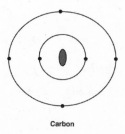

Carbon

Figure 16: A carbon atom with its four outer electrons.

In order to understand the ways in which the elements contribute to Gaia we need to learn a further rule of atomic social etiquette, namely that rather than sharing electrons in a covalent bond, it is possible for an atom to give away one or more outer electrons if there is an atom available that can use them to complete its own outer orbit. This type of atomic union is known to chemists as an *ionic* bond, and it happens when the donor atom tends to have less than four outer electrons. On the other hand, says the rule, if an atom has more than four outer electrons, it will tend to accept additional electrons from other atoms when bonding ionically. A good example is common salt, sodium chloride. The sodium atom has a single outer electron, which it happily gives to a chloride atom that needs only one electron to complete its outer orbit. By exchanging the electron both partners find rest, but now the attractions between positive and negative charges come into play. Having lost an electron, the sodium atom now has one proton in its nucleus without an equivalent negative charge to balance it, so the sodium atom has become a being that attracts negative charges. Meanwhile, the chlorine atom, in taking up the electron, has become a chloride ion and now has an overall negative charge. The sodium and chloride, as charged atoms (which chemists call *ions*), experience a new restlessness—they must lie close to each other so that their charges can interact, an attraction which manifests, under the right conditions, as the crystals familiar to us as table salt.

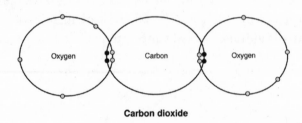

Carbon dioxide

Figure 17: Mutual fulfilment for carbon and oxygen atoms
in the covalent bonds in the carbon dioxide molecule.

Hydrogen (Figure 18) is the most abundant atom in the universe: 88% of all atoms are hydrogen, and indeed hydrogen is the primordial atom from which all others are derived through fusion in the intense heat and pressure within stars and supernovae explosions. It the simplest and lightest of all atoms: in its most basic form it has just one proton in its nucleus, which gives it its gravity-defying lightness, and one electron in its single orbit. Hydrogen ions are the most chemically reactive ions in existence, and also the smallest. Hydrogen seeks fulfilment by finding another electron for its single orbit, which holds only two. This means that two hydrogen atoms happily bond to each other covalently to make H_2—a hydrogen molecule—but hydrogen also bonds cheerfully with other elemental beings such as carbon, phosphorus or oxygen, and is a major constituent of living beings. Hydrogen is an airy, flippant creature which would love nothing better than to escape our planet altogether and return to its ancestral domain in outer space as hydrogen gas, for Gaia's gravitational field is not strong enough to keep it from floating off into the vastness of the cosmos. If this were to happen, we would lose all our water and the planet would dry out completely. This may well be what happened on Venus, but life on our planet has various ways of recapturing free hydrogen by combining it with oxygen before it can escape.

Hydrogen

Figure 18: A hydrogen atom with its lone electron.

Nitrogen encounters its most stable relationship with itself. It has a full inner orbit, but misses three outer electrons. This means that two nitrogen atoms create a triple covalent bond which requires a great deal of energy to break apart; two nitrogen atoms linked in this way are like virtually inseparable twins, and nitrogen gas is thus highly unreactive. In a process invented by the German chemist Haber, nitrogen gas is forced to react with hydrogen to make ammonia, but this can only be done with a catalyst present and at temperatures of 500^0C and a pressure 1,000 times greater than normal atmospheric pressure. Perhaps nitrogen suffers when it is forced to separate under these extreme conditions; it may feel more comfortable when bacteria carry out this operation, for they are the only life-forms that can bring about this extraordinary feat at everyday temperatures. Working in partnership with plants such as clovers and legumes, which house the microbes in special root nodules, they split the nitrogen shielded from the ardent and potentially destructive attentions of oxygen. It is essential that nitrogen does indeed stay twinned as molecular nitrogen gas (N_2), for, as the most abundant gas in the atmosphere (78%), its collective weight at the Earth's surface produces the right pressure for the greenhouse effect which regulates the Earth's temperature, although any extra nitrogen would asphyxiate oxygen-breathing life. Atmospheric nitrogen also dilutes oxygen, which is thereby restrained from consuming anything in its path in spectacular global combustion. Nitrogen is a nutrient essential to all life, and is of critical importance for the formation of DNA, the haemoglobin in our blood and amino acids, which can link up thanks to nitrogen, forming protein molecules of potentially vast size. Nitrogen is of course also the basis of dynamite and TNT. These compounds are so famously explosive because nitrogen will combine with itself at the slightest opportunity, releasing vast amounts of energy in the process.

Oxygen (Figure 19) is the third most abundant element in the universe after hydrogen and helium, but is the most abundant element in the Earth's crust. In very big stars—far bigger than our own sun—oxygen nuclei can fuse to give rise to silicon, phosphorus and sulphur, all of which have more protons than oxygen, and also an extra electron orbit to balance the extra positive charge. Having a full inner orbit but only six electrons in its outer shell makes oxygen passionately hungry for electrons—so hungry in fact that it can find fulfilment by bonding covalently with virtually every single known element. Only helium, neon, argon and krypton are immune from its fiery attentions, because these 'noble gases' enjoy complete outer orbits and are hence serenely aloof from the hurly-burly of the everyday soap opera of chemical life. Oxygen atoms love to bond covalently with each other, but in so doing a curious chemical anomaly becomes apparent—two electrons from each outer orbit refuse to join the melée, and remain unpaired. When oxygen is liquefied at very low temperatures, these free electrons make oxygen an excellent conductor of electricity. Oxygen is the passionate Italian of the chemical world—its urge to gather electrons is so powerful that it can literally burn up the complex molecules of life, releasing copious quantities of solar energy originally locked up by photosynthesis. Respiration, without which multicellular life, such as us, would be impossible, uses oxygen to burn up food molecules in a gradual, controlled way and stores the energy in special molecules such as phosphorus-rich ATP. As we saw earlier, abundant oxygen, together with combustible gases such as methane in a planet's atmosphere, are a sure sign that life is present, for only life can release vast quantities of these gases into the air.

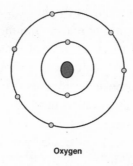

Oxygen

Figure 19: An oxygen atom with its six outer electrons.

So passionate is oxygen in its quest for electrons that once inside the cell its reactions give rise to highly toxic free radicals that can interfere with DNA, causing ageing and even cancer. Free radicals are atoms that end up with a missing electron when the weak bonds that they have been involved in break up, especially after they have experienced the ardent attentions of oxygen. The free radicals then feel impelled to capture electrons from neighbouring molecules, thereby creating new free radicals that oftentimes set off a potentially highly damaging cascade effect. Cells have invented a host of enzymes to mop up the rogue chemical beings, but a small number evade capture to carry out the demolition of the genetic material—a process that will eventually kill most of us. So, like all the nutrients that are essential for life and even solar energy itself, oxygen is both a life-giver and a dealer of death.

Phosphorus as a lone element is never encountered in nature. It was first collected in its pure form by Henning Brandt in 1669 from urine, perhaps his own, which he evaporated to leave a residue that he heated until red-hot. He collected and condensed the resulting vapour, ending up with a white powder that glowed in the dark and spontaneously burst into flame when exposed to the air. These properties give phosphorus its name, from the Greek words *phos* (light) and *phorus* (bearing); hence it is the 'light-bearer' of the chemical world. In living beings, phosphorus is involved in light-bearing in more subtle ways: firstly through its central involvement in the storage and release of energy, ultimately derived from the sun, that powers the light of sentience in life itself; and secondly as the ultimate source of the eerie light of bioluminescence so much favoured by some planktic algae and by creatures of the ocean deeps. Some speak of phosphorus as Gaia's 'master nutrient' because of its scarcity (it is the eleventh most abundant element in the Earth's crust) and because of the difficulty of returning it from the marine sediments, where it is relatively abundant, to the land and to the surface of the ocean, where it is scarce; this feat is made especially difficult because there is no gaseous molecule in wild nature that acts as the agent for this transfer.

Phosphorus has its two inner orbits nicely replete with electrons, but its outer orbit, the third one out, misses three. Phosphorus links with carbon and nitrogen to make the ATP molecule, which is the front-line energy acceptor molecule present in every living being, whether microbe or elephant. We humans process an astonishing one kilogram of ATP per hour, every day of our lives. It is also found in the DNA molecule, and in vertebrates occurs most commonly in bone, linked to calcium and oxygen atoms as calcium phosphate. Like oxygen, phosphorus is a passionate chemical being. When the two naked elements encountered each other in the air, the flames that shocked Brandt bore witness to their ardent and explosive attraction to each other. The result of their fiery affair is the phosphate ion, in which phosphorus is linked to four oxygen atoms.

Sulphur, like oxygen, has a third, outer orbit, which misses two electrons. To arrive at sulphur, take oxygen, fill its outer, second orbit, and add a third, but with two electrons missing. Sulphur, like oxygen, thus seeks two electrons for completion, but in a less frenetic way; it has a mellower (and yellower) character because its two inner orbits are satisfactorily full. It is essential for life because it is a key component of the amino acid methionine, without which human biochemistry cannot be sustained. Sulphur shares some of the passion of oxygen, but being more moderate in its relationships, is able to form long rings and chains rather like those of carbon, and to bond with other atoms in a great variety of ways. Strangely, sulphur in its various gaseous forms it is often associated with smells that are offensive to the human nose.

Many other elements are crucial for Gaia, and of those calcium, iron and silicon are of great importance. A calcium atom has four electron orbits, three of which are full, but its outer orbit has only two electrons, which makes it only too happy to engage in ionic bonds by giving these electrons away, leaving behind a calcium ion which carries a double positive charge. It is this that makes calcium so attractive to negatively charged ions. Calcium has been called the messenger of the cell, because it is somewhat like those charismatic entrepreneurs of the human world who have a mercurial aptitude for networking. It is involved in virtually every cellular process, ranging from cell division to fertilisation to muscle contraction, and without this Hermes-like atom the astounding coordination of the cell's metabolism would be impossible. But too much calcium can kill, so cells must expend energy to keep it at a concentration low enough for optimum functioning.

Iron, like calcium, has three full electron orbits, but has eight outer electrons rather than two. This means that iron needs to seek ten electrons to find completion, and it does this in a variety of ways. It has a particular penchant for oxygen: iron sits at the centre of the haemoglobin molecule, where it binds with oxygen from the lungs, releasing it to cells where the oxygen tension is low. It also binds to oxygen in two important forms: as haematite (Fe_2O_3), familiar in black volcanic sand beaches, and magnetite (Fe_3O_4). Both of these iron compounds had an important part to play in determining the oxygen concentration in the Earth's early atmosphere.

Silicon, like carbon, needs four electrons to complete its outer orbit, but unlike carbon it has an extra inner orbit with eight electrons. Silicon shares some of the character of carbon—it is a highly social being which likes to make long chains with other silicon atoms, and is very fond of doing so as the silicate ion $(SiO_4)^{-4}$ in which a silicon atom links up with four oxygen atoms. Silicate ions can link up in an astonishingly large number of ways, one of which occurs when the four oxygen atoms are shared, giving rise to silica (SiO_2), which can arrange itself into the highly ordered, spiral configuration of quartz crystals. The continents and the sea floor are made out of calcium-silicate rocks which can take many complex forms but basically consist of a silicon-oxygen scaffolding with many other atoms, including ionic calcium, held within it by the ubiquitous forces of electrical attraction. The continents are made out of granite, and the sea floor out of basalt, both of which, as we shall see later, are crucially important in regulating Gaia's surface temperature.

The Origin of the Elements

We have now learnt a few basic things about the personalities of the major elemental beings crucial to Gaia, but where did they come from? This question is relevant because, by taking us into the realm of cosmology, it helps us to understand not only the origins of our planet, but also to ponder its existence in relation to the universe as a self-organising, evolving entity. One answer to the question is: 'out of the Big Bang'. In this mysterious event, which happened some 15,000 million years ago, energy, matter, space and time appeared out of nowhere in a primordial instant of creation. The startling thing about this is that the universe didn't appear suddenly into a pre-existing space because space was itself created along with the primordial energy, which we can still detect as the 'cosmic background radiation.' As the fireball spread it gradually cooled, until after the first fifteen minutes the energy condensed into first electrons, and then, with further cooling, into neutrons and protons. The nascent universe spread out as new space came into existence in between these newly formed elementary particles, and eventually, in the next fifteen minutes, the cooling was sufficient to allow the coalescing of these particles into hydrogen, the first elemental being born of the universe, and to this day the most abundant of all the elements.

The Big Bang produced none of the heavier elements, other than some helium and lithium, hydrogen's immediate neighbours in the periodic table. The heavier elements were forged much later on, when atoms of hydrogen clumped together through gravitational attraction to form the stars that coalesced as the universe cooled even further. Eventually some of these clumps became so big that the pressures at their centres were great enough to fuse hydrogen nuclei together into helium, releasing immense amount of energy, some of it as visible light. In this way stars were born, and the whole universe lit up with their brilliance. The astonishing thing is that the conditions that lead to the birth of stars are very finely balanced. If the gravitational attraction between bits of matter had been just a little greater or smaller than it actually is, stars as we know them would not have been possible, and without them the elements, the solar system and the life our planet bears could not have existed.

For the first few billion years the universe was filled only with stars that burnt hydrogen, but as some of the largest stars aged into red giants during the last 10% of their lifetimes, heavier elements were formed in their immensely dense, hot centres. For much of their life these stars burnt hydrogen, but then, as they died, pressures and temperatures reached such high levels in their interiors that heavier elements were formed at an ever-quickening pace. Carbon atoms were created when groups of three helium nuclei fused, giving carbon nuclei with six protons and six neutrons, accompanied by vast releases of energy. Then, as the stars aged even more, the denser elements such as sodium, magnesium and oxygen were born as some of the carbon nuclei fused, until the sequence reached iron, after which the creation of new elements stopped in all but the very largest stars. Once the iron phase was reached inside these stellar giants, the pace of elemental creation was frenetic, and in the last few seconds of their lives an inward gravitational collapse generated sufficient energy to power massive supernova explosions which sent vast quantities of hydrogen and smaller amounts of the heavier chemical beings such as carbon, oxygen, phosphorus and sulphur whirling into the outer reaches of interstellar space. Some of these clouds of atomic beings coalesced into new stars, giving rise to fresh supernova explosions and more newly synthesised elements—a process that goes on to this day. Thus many of the chemical beings which now constitute the Earth, and indeed our whole solar system, have lived in several stars before coming to dwell in us and in the rocks, atmosphere and ocean of our planet.

The Birth of Our Solar System

Sometimes a cloud of material flared forth by a supernova condenses into a nebular cloud of dust, and this can in time differentiate out into a solar system. One such cloud, in the outer reaches of a galaxy known as the Milky Way, became the solar system in which Gaia resides. The cloud of interstellar matter that eventually became our Earth had just the right combination of elements to give rise to a living planet. An astounding 99% of the mass of the nebular cloud was hydrogen, with heavier elements weighing in at a mere 1%. However, amongst these relatively rare chemical beings there were sufficient radioactive materials, such as radioactive potassium, uranium and thorium, to provide a sufficient energy source in the inner earth to drive the movements of the Earth's tectonic plates; there were also carbon, oxygen, hydrogen, sulphur and nitrogen, which are essential for life, plus other elements such as iron, calcium magnesium and oxygen, which formed the Earth's crust and its deep rocky interior.

At the centre of the nebular cloud was a proto-sun, and around this was a flattened disc of interstellar dust which gradually organised itself into bands, and then into small grains of sand within each band as more and more dust particles collided. In any given band these tiny grains of sand gradually got together by means of gravitational attraction and by simply bumping into each other to make larger and larger clumps of matter, until eventually all the matter had been swept into a planet, which was left orbiting the trajectory where each band of dust had once been. All the planets in our solar system came together in the same way, but the first four, Mercury, Venus, Earth and Mars are rocky; the next four (Jupiter, Saturn, Uranus and Neptune) are gaseous. Pluto, the last, is also mostly rock.

It was fortunate that the nebular cloud that eventually formed the Earth was poor in carbon and water, for too much carbon would have meant so much carbon dioxide in the atmosphere that the surface temperature of the Earth would have been very high right from the outset. The high temperatures would have evaporated all of the abundant water into the atmosphere, driving the temperature still higher because water vapour is a powerful greenhouse gas in its own right. With all its water in the atmosphere, Earth would have faced the inevitability of total desiccation. High-energy sunlight at the top of the atmosphere would sunder the bonds between hydrogen and oxygen like so many swords cutting through butter, and the newly liberated hydrogen atoms would have then eagerly shot off into space because hydrogen (a being with basically just one proton and one electron) is so light that Earth's gravity cannot prevent its escape into the surrounding void. The remaining oxygen, bereft of hydrogen, would have satisfied its hunger for electrons by reacting with sulphur gases from volcanoes to give sulphuric acid. With no hydrogen at all and no free oxygen, the water could never have been reconstituted. This was the fate suffered by Venus, our nearest sunward neighbour, where the surface temperature is high enough to melt lead and the acidity is sufficient to corrode even the most resistant of materials. But things on our planet were different: its initially low levels of carbon and water were supplemented to just the right extent by impacts with meteorites and comets from the outer reaches of the solar system where these elements were much more abundant.

Our infant planet was special in many other ways. Its orbit was just the right distance from the sun to allow liquid water to remain on its surface, and its mass provided just the right amount of gravitational attraction for holding the atmosphere and ocean in a protective embrace around the Earth. The sun itself provided a relatively steady output of energy, without too much sterilising ultraviolet radiation. The configuration and masses of the other planets in the solar system were also well tuned, so that mutual gravitational influences on each other and on the Earth produced the enduring emergent property of stability in the Earth's orbit. Had any of the masses of the planets been even slightly different, our own planet's orbit may well have been chaotic, making the evolution of a Gaia with complex multicellular life impossible. The moon was created when a Mars-sized piece of interplanetary shrapnel struck the Earth some 4,500 million years ago. The impact released so much energy that both planets melted to their cores, sending a great mass of molten debris into space which later condensed into our nearest planetary neighbour. Our moon is critically important for the living complexity of our planet, for her intense gravitational embrace further stabilises our planet's axial tilt, which would otherwise wobble chaotically. The gravitational field of a large Jupiter-sized planet is also needed to deflect asteroids—large chunks of interstellar rock—away from the Earth, although Jupiter's immense gravity does occasionally send asteroids and comets in our direction. In the very distant past, about 4,000 million years ago when our planet was forming, some of these comets (which were then far more abundant) may have provided Earth with a much-needed source of water.

From all this you can see that it took more to make Gaia than the mere presence of the right chemical beings in the right proportions in a condensing protoplanet, for the wider context of the solar system was also important. But even this was not enough: the galactic context was also crucial. Gaia has evolved in the Milky Way, a galaxy with enough heavy elements and with the right shape to support a viable solar system. Furthermore, Gaia lives in just the right *part* of the galaxy, safe from the sterilising gamma rays that emanate when super-massive stars collapse. It is as if matter was waiting for the appearance of the right conditions before it could explore the possibilities latent within itself for the emergence of an evolving, self-regulating planet hosting an abundance of life. Matter ached to experience itself unfolding into the fullness of the living state. Was it pure chance that the right conditions appeared so long ago in our part of the galaxy? Could it be that there is a difference between chance and luck? Chance is deterministic, perhaps, whereas 'luck' could imply the action of an indeterminate animating principle within all things. If so, luck had played its part in setting up the right conditions, and the spinning ball of rock that was to become Gaia was now ready to take her next step: the appearance of life.

Plunging into Deep Time

Find somewhere you can relax and make yourself comfortable, perhaps your Gaia place. Take a few deep breaths, and when you are ready imagine that you are under a starlit sky standing at the edge a well, the well of deep time.

Peering into the well, you see a universe of stars swirling in an inviting inky space much like that of the heavens above you.

Feeling an intense curiosity, you walk onto the edge of the well, and ease yourself into its centre with a complete sense of safety and confidence. You float gently in the enfolding embrace of the well's inky medium.

Slowly, you float down a little way, and suddenly you see Gaia from space as she was just before the industrial revolution, with her swirling mantle of white clouds, her ample blue ocean and her continents spread far and wide on the surface of her lustrous spherical body. Great biomes cover the land with rich, diverse vegetation. As a bird, you fly through each biome in turn, sensing the massive abundance of living beings that dwell in each one. As a fish, you plunge into the ocean and sense the amazing diversity of life in that great watery realm.

A knowing comes to you that Gaia now hosts more species then ever before. You sense how sublimely sophisticated she is at handling the chemical elements and wonder at her inordinate skill at dealing with the brightening sun. You notice how the webs of relationship that weave together her life forms and her rocks, atmosphere and water are more tightly coupled now than ever before. It is clear to you that Gaia has reached the pinnacle of her evolution.

Now plunge deeper into the well of time, back 100 million years ago, to the time of the Age of Reptiles. Once again you see Gaia from space, but now her continents are closer together, and the Atlantic

Ocean is much smaller. Diving into the vegetation, you are stunned to see how it has changed. There are giant tree ferns with immense centipede-like leaves, tall cycads and primitive flowering plants. In amongst the tree ferns are giant dinosaurs moving around with unexpected ease, and on the ground you spy familiar-looking ants and termites. In the air, small dragonflies dash and race.

Plunging deeper into the well of time, you reach 250 million years ago—the Age of Amphibians. From space, you see only a single great continent straddling one side of Gaia's face, reaching to the high latitudes on either side of the equator, lush with plant growth. Now there are tall palm-like trees, and tall spiky horsetails and seed ferns. You spy huge salamander-like amphibians and primitive reptiles that lay leathery eggs in the warm sand of the ample river beaches. Giant dragonflies dart through the air, and cockroaches scuttle through the decaying vegetation.

You dive deeper into the well's velvety embrace, enjoying the sensation of sinking into time, like falling asleep into a rich and multifaceted dream, until you reach 500 million years ago when some of Gaia's first multicellular beings made their appearance. You notice how different she is now when seen from space. The continents are now clustering together, and gone is the lush vegetation on land, for now only a thin smearing of microbial crusts and mats covers the rocks. Plunging into the shallow sea at the edge of the continent you see a host of armoured trilobites, sea scorpions and strange spiral-armed starfish. Sponges sift the nutritious seawater, and microscopic silica-shelled radiolarians hunt for phytoplankton prey by extending their bulging pseudopods into the light-filled sea.

You take your leave of this almost familiar world, and plunge deeper and deeper into time, falling now like a slowly twirling autumn leaf for much longer than before, until you reach 1,800 million years ago. The smaller continents are still covered with rock-dissolving bacterial mats, and there are large red deposits on the land, indicating that free oxygen is still present in the air. Diving into the shallow seas on the continental shelves, at first you see no life at all. Shrinking yourself down to the size of a full stop, and then one thousand times smaller

*still, you encounter a multitude of tiny single-celled beings, each con-
taining a spherical globule, the nucleus, which holds their genetic
material. You also see a host of even tinier sausage- and corkscrew-
shaped beings—the bacteria.*

*Staying small, you plunge down even further to 3,500 million years
ago. Seen from space, Gaia is no longer a blue jewel, for now there
is precious little oxygen in the atmosphere. The sea is greenish,
reflecting a pinkish methane-dominated sky. Gone are the abundant
continents; you see only a smattering of volcanic islands dotted
about in the sea, into which you fall like the tiniest mote of dust.
There in the upper sunlit reaches of the water you spy only bacteria—gone are the larger nucleated beings of the earlier world. Some
of the bacteria are green, and exude small bubbles of gas—oxygen—
which is quickly gobbled up by oxygen-hungry iron and sulphur
compounds. Large bacterial colonies, the stromatolites, secrete their
chalky domes wherever the waters are shallow enough to support
them.*

*Now you fall into the deepest recesses of time, until you reach 4,600
million years ago, long before Gaia existed, when the Earth was a
newly formed ball of rock. Floating in space, you see huge comets and
meteorites bombarding the nascent planet, provisioning it with water
and other key ingredients for her future as the mother of life. Astonished, you watch as a massive planet the size of nearby Mars crashes
into the Earth, melting both partners and sending shards of molten
rocky debris into orbit that eventually coalesce into the moon.*

*Melodious bird song from the upper reaches of the well breaks the
spell, calling you back to our own times. Quickly you swim upwards
through the blue-black velvety fabric of time to the mouth of the well,
and return to the broad night sky of the present moment.*

Chapter 5

Carbon Journeys

Carbon in the Long Term

We've already examined some of the evidence for Gaia—remember the geologist's trace of her temperature, which has remained within habitable limits over the 3,500 million years of life's tenure on our planet, despite occasional excursions into hotter and colder periods. Remember too that an ever-brightening sun, combined with continuous emissions of carbon dioxide from volcanoes, should have resulted in a hellish, life-obliterating global super-hothouse many millions of years ago; but yet we enjoy a fairly comfortable global mean temperature today of about 15^0C. Why? The answer involves a wonderfully holistic and delightfully Gaian combination of biology (life), geology (rocks), physics (energy transfers) and chemistry (interactions amongst the chemical beings), working together to regulate Gaia's temperature over a range of time scales in a never-ending dance of negative feedback.

Let's look at the story of how this happens in the very long term, over a million years or so. One can recount this story either in the amazingly dry language of conventional science, in which everything is treated as if it were just dead matter observed from afar by a vastly aloof human intellect, or one can tell it by acknowledging our inescapable embeddedness in Gaia, and our intimate connection to the animate qualities within every speck of matter. Objectifying dryness utterly dominates conventional scientific writing in both popular and technical genres, so I am going to draw on an unashamedly animistic version of this (and other) Gaian stories by deliberately using personifying as a device to help breathe life back into what might otherwise be a rather boring account, capable of exciting the imaginations of no more than a few handfuls of ivory-towered specialists.

So let me begin at the ultra-microscopic scale by introducing you to *Emiliania huxleyii* (Figure 20), a single-celled marine alga that lives at the surface of the cold oceans as a member of the phytoplankton community. *Emiliania* is tiny, 4/1000th of a millimetre (4 microns) in diameter, so the wheel-shaped structures you see in the picture are even tinier—you need an electron microscope to see the fine detail revealed in the photograph.

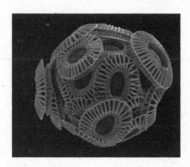

Figure 20: The coccolithophore *Emiliania huxleyii*, with its chalky plates.
(*photo © Steve Gschmeissner / Science Photo Library*)

Emiliania belongs to a group of algae that revel in a delightfully romantic name: they are *coccolithophores*, meaning the 'carriers of little stone berries'. The berries in question, the wheel-shaped structures or coccoliths, are made out of one of Gaia's most important molecular beings: calcium carbonate, a combination of three of the elements born of the supernova explosion which led to Gaia (calcium, carbon and oxygen), the most common form of which is *calcite*. Calcite can itself manifests in a variety of ways, but is most commonly encountered as light porous chalk, or as the much denser limestone. *Emiliania* is a single-celled photosynthesiser, and is a master of using the energy of the sun to convert carbon dioxide and water into sugars and oxygen. But she is also skilled in another complex biochemical art—the precipitation of calcite within deep intracellular chambers into exquisitely crafted coccoliths, which, when complete, are excreted to surround the cell in a white coating of chalky plates. Chemists write down the formula for calcium carbonate (the chalky stone berries carried by these little creatures) as follows:

$$CaCO_3$$

This means that a single molecule of chalk contains one atom of calcium, one atom of carbon and three atoms of oxygen. As we shall soon discover, it is the presence of the carbon atom which gives chalk its significance for Gaia's long-term temperature regulation.

Now let's move from the micro-scale of the story to the far vaster spatial scale involving movements of the Earth's crust and the immensely powerful activity of volcanoes, which spew out carbon dioxide—the greenhouse gas—along with large quantities of lava, a kind of rock closely related to basalt and granite. Basalt is the mother of all rocks. It wells up at Gaia's mid-oceanic ridges, hot and pliable like just melting chocolate from deep inside the Earth. Granite is born at super-high temperatures and pressures deep below the continental margins when basalt is cooked with water, or when calcite and silica deposits re-combine. Basalt and granite (known to scientists as calcium silicate rocks) contain a lot of calcium, oxygen and silicon which self-organise on cooling into three-dimensional crystalline lattices of exquisite precision and regularity. Locked up in the rock lattice like captive princesses in an ancient castle are positively charged calcium ions which, never losing hope of experiencing something other than the stasis of a crystalline existence, long to escape the lattice prison that has held them captive for often millions of years. There is only one way that calcium's escape can be assured—she must embark on a chemical marriage with carbon, her prince, suitor and bridegroom who, riding the atmosphere as part of a carbon dioxide molecule, searches everywhere for his rock-incarcerated princess. When a carbon dioxide molecule finally encounters basalt or granite the marriage can at last happen, but only after some complex challenges have been overcome.

First, a water molecule from a rain shower must dissolve a carbon dioxide molecule to yield carbonic acid that immediately dissociates into two new chemical beings: a bicarbonate ion, in which the carbon atom is linked to one hydrogen and three oxygen atoms, and a positively charged hydrogen ion. The hydrogen ions thus released, being nothing more than protons, are small enough to travel easily amongst the much larger chemical beings such as calcium, silicon and oxygen which hold the rock lattice tightly together because of the way in which their positive and negative charges interact and self-organise. Vast hordes of the tiny, positively charged hydrogen ions insinuate themselves into the rock. Ant-like, they pass through tiny gaps in the walls of the granite castle and cluster around the negatively charged oxygen and silicon ions, neutralising the electrical attractions that hold the rock together, so that what was once solid, impassable granite or basalt slowly dissolves like a wet sugar lump.

As the rock falls apart, carbon from the atmosphere, held fast in the bicarbonate ions, combines with the newly liberated princesses, the calcium ions, merging in sacred chemical marriage to become *calcium bicarbonate*, which loosely speaking is a water-soluble form of chalk holding within it carbon dioxide removed from the air. There is a precise calculus here: each calcium ion weathered from the rock links up with two carbon atoms from the atmosphere. The celebrants in these chemical marriages are so eager to combine with each other that even a single rain drop falling on the bare surface of the smallest pebble of basalt or granite is sufficient to initiate many chemical unions, as calcium ions are at last released from their rocky prisons to conjoin with their ardent carbon suitors plucked from the atmosphere. This process, known to mainstream science as *calcium-silicate weathering*, removes carbon dioxide from the air, thereby cooling the Earth.

The calcium bicarbonate solution gets flushed through the soil by rainfall, and eventually finds its way into rivers that carry it to the sea, where, if they are present, the coccolithophores precipitate it as solid chalk within their microscopic bodies. Other marine creatures also precipitate chalky shells and body armour from calcium bicarbonate, amongst them crustaceans such as crabs and barnacles, and some molluscs, such as clams, oysters and cuttlefish, whose lozenge-shaped chalk cuttlebones can often be found washed up on beaches all over the world.

By precipitating chalk containing carbon dioxide stripped from the atmosphere, these beings have a massive cooling effect on the entire planet. But it is the microscopic floating chalk-forming phytoplankton that have provided the lion's share of this cooling effect over the last 80 million years or so. These tiny algae thrive over vast areas of the cold oceans of the world. When they die, a chalky 'marine snow' settles on the bottom of the sea, squeezing and squashing underlying accumulations of chalky skeletons into solid chalk rock. In many parts of the world these chalk deposits can be seen above sea level where they have been uplifted by the great movements of Gaia's crust. The famous chalk cliffs known as the Seven Sisters in southern England are a marvellous example. These cliffs are made almost entirely out of countless numbers of microscopic chalky skeletons secreted mostly by coccolithophores. When you look at these cliffs, or indeed at any other chalk or limestone, you are of course seeing rock, but seen through Gaian eyes, what you are actually looking at is atmosphere made solid, or more specifically, carbon dioxide distilled and solidified out of the atmosphere thanks to the irresistible attractions between carbon dioxide and the calcium princesses that dwell in the very heart of the calcium-silicate rocks.

But these great deposits contain more than chalk. Even a casual look at the rocks of the Seven Sisters will reveal a large number of hard nodules of flint that resist weathering and are often transported over large distances by tides and currents. These flints are made of the silica and oxygen weathered out of granite and basalt at the same time as calcium. Washed into the rivers as silicic acid, the silica reaches the sea, where diatoms (Figure 21), radiolarians and sponges precipitate it into exquisitely crafted glassy shells and spicules that rain down to the murky depths of the ocean alongside the chalk shells of the coccolithophores. How the minute glassy remains of these beings coalesce into flint nodules is still somewhat of a mystery.

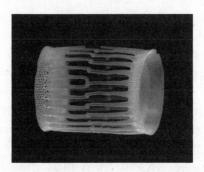

Figure 21: A diatom with its finely crafted silica exoskeleton.
(*photo © Steve Gschmeissner / Science Photo Library*)

Calcium bicarbonate from rock weathering is so plentiful in the oceans that marine creatures face the very real danger of death by calcite encrustation, for calcium bicarbonate likes nothing better than to transmute itself into calcium carbonate by precipitation onto available surfaces—just as it does in limestone caves. There is a deep comfort for each atom of carbon, oxygen, hydrogen and calcium when they configure themselves into calcium carbonate—a great sense of reassurance and stability for each of them in the presence of the others. But these cosy familial embraces amongst the chemical beings are both a blessing and a curse as far as living things are concerned: a blessing because calcium is essential for the inner life of the cell, and because it can be used to make tough protective armoured plates; but a curse because too much calcite spells death through encrustation, or through poisoning of the cell's inner workings. Just about every single creature that lives in the sea deals with these dangers by expending a lot of energy in pumping excess calcium out of its cells, and by producing complex slimes replete with specially crafted sugar molecules that stop the encrustation in its tracks. Some marine organisms simply exude these chemicals onto their skins and into the surrounding sea water, but others, like the coccolithophores, crabs and sea urchins, have mastered the art of controlling the precipitation within intracellular spaces in which the special sugary slimes are used to direct the manufacture of a vast array of intricately sculpted calcium carbonate artefacts such as carapaces, wheels, spicules, rods and stirrups. Even land animals secrete these anti-calcification slimes—we lay them down on our teeth to keep them clear of encrusting calcium phosphate. So replete is the ocean with dissolved calcium bicarbonate from the weathering of rocks that every single surface would become rapidly encrusted with calcite if the sugary anti-calcifying slimes were somehow removed.

The control of calcification by living beings has become ever more subtle and sophisticated as Gaia's ability for self-regulation has evolved and strengthened over geological time. The most basic kind of calcification happens without the intervention of life when, in the absence of any chemical inhibitors, calcium precipitates out of the water to form simple marine crusts or the stalactites and stalagmites in limestone caves. But in the heyday of Gaia's infancy, some 3,000 million years ago, huge bacterial communities (the stromatolites) laid down vast crusty platforms of limestone wherever they needed to grow nearer to the life-giving light at the ocean surface. Then, from 600 million years ago until about 80 million years ago, the main calcifying beings were the corals, molluscs and crustaceans. Since then the main site of life-enhanced calcification has shifted from the margins of the continents to the outer edges of the continental shelves, as an infinitude of tiny floating beings such as the coccolithophores have laid their chalky dead to rest on the seabed far below them. During this whole evolutionary trajectory from crusts to coccolithophores, Gaia has become increasingly skilled at calcification, so that today far greater quantities of chalk are precipitated out of the ocean by microscopic algae than ever before, a development that has allowed Gaia to keep herself cool in the face of a dangerously bright sun. Once again, Gaia reveals herself as an *evolving* animate entity.

So far we have only considered the role of ocean life in Gaia's great chalk journey, but life on land also makes a vitally important contribution to cooling the earth by greatly enhancing the physical and chemical dissolution of basalt and granite. Wherever life on land grows on these rocks, it uses a variety of physical and chemical means to extract the rich lode of nutrients they bear. In so doing life accelerates and enhances the chemical entanglement of water, carbon dioxide and calcium into calcium bicarbonate, a liquid trap for carbon dioxide that is eventually deposited at the bottom of the sea by the chalk-forming sea creatures, and it also speeds up the release of silica that ends up in the glassy bodies of the diatoms, sponges and radiolarians.

How does life on land accomplish this enhanced weathering? You are a calcium ion locked away in the crystal lattice of a huge granite dome, and you have been sitting here, some ten metres from the surface, for tens of millions of years. As you look through the lattice you see a great deal of empty space, but you spy other calcium ions, and indeed other chemical beings such as oxygen, silicon and aluminium, all in regular crystalline arrays. It's like putting your head between two parallel mirrors in a Chinese restaurant where there are a few white lanterns above you. Look into either mirror and you see serried ranks of lanterns curving off to seeming infinity on either side. The white lanterns are your fellow calcium ions. You would need to wait for a very long time to escape if you had only chemical rock weathering to rescue you. But the rock surface ten metres above you is in fact teeming with countless living beings in a rich, dark soil which supports fungi, microbes and large plants such as trees and shrubs. A massive tree root snakes its way towards you through a natural joint in the granite. The root physically splits open the rock, and in its wake comes rich, black, moist soil, full of microbial life. Some of these microbes are bacteria which secrete a complex sugar molecule that swells when wet, splitting off little grains of granite from the main rock surfaces of the tree root tunnel. These little granite fragments provide a massively increased surface area for the weathering, and are bathed in the gaseous exhalations of billions of teeming microbes that populate the soil. These microbes, like us, use oxygen from the air to extract energy from food molecules and breathe out carbon dioxide. As a result the soil is far richer in this gas than is the air above. The tree roots themselves exhale carbon dioxide as they use oxygen to burn sugars made by photosynthesis in leaves way up in the churning atmosphere. Furthermore, the soil is made wonderfully porous thanks to the actions of many small creatures such as woodlice, millipedes and earthworms, who turn it over like so many gardeners, allowing rainwater to easily percolate down to complete the chemical marriages between calcium and carbon dioxide on the tiny fragments of rock.

Life on land is a great rock-crushing, rock-dissolving being. Weathering can be considerably enhanced by even relatively simple beings such as lichens and bacteria growing on the rock surface, but trees and shrubs can reach deeper into the rock, making the whole process happen very much more quickly. Thus in warm wet tropical conditions, life can accelerate granite and basalt weathering as much as 1,000 times relative to a bare, lifeless surface. If you are a calcium ion in tropical rain forest granite, you might wait only one year rather than a thousand to begin your sea-bound marriage with your two bicarbonate ion partners.

But there is a great danger here: the weathering of granite and basalt could send the planet into a permanently frozen ice-ball state if too much carbon dioxide from the atmosphere drains away as chalk to the bottom of the ocean. So why hasn't this happened? Gaia as a whole prevents this icy fate thanks to the tectonic movements of her crust, driven by the great powers residing in the deep interior of her rocky body. Here there is a great deal of radioactive material left over from the supernova explosion that gave rise to the elements of our planet. When these radioactive materials decay, a vast amount of heat passes into the surrounding semi-molten basalt rock, which rises in great plumes like so much hot air until it reaches the mountainous sub-sea spines, some of which run north-south in the middle of the Atlantic and Pacific oceans. Here the newly emerged basalt cools and spreads away from the ridges as new rock wells up behind it. All the while chalky bodies settle on these great moving plates of sea floor basalt, as they journey slowly but surely towards a continent. In many places, the bodies of the silica-shelled beings—the diatoms and radiolarians—settle on top of the chalky corpses, protecting them from the chalk-dissolving powers of the acidic, high-pressure deep ocean waters. The continents themselves are mere passengers on these huge moving slabs of basalt, for their granitic foundations are lighter than the underlying plates that carry them about the world.

When it finally meets a continental edge, the sea floor basalt gently bends beneath it, carrying some of the overlying calcium carbonate and silica with it into the Earth's depths, much as a great diving whale carries down the barnacles which pepper its thick blubbery skin. As the calcium carbonate, silica and basalt plunge into the abyssal deeps, they melt under temperatures and pressures powerful enough to break the bonds between calcium and carbon in the chalk, and between silicon and oxygen in the silica. And then two extraordinary transformations happen. The first is the release of carbon dioxide that rises upwards beneath the edge of the continent, breaking through at last in spectacular volcanic eruptions that return vast amounts of carbon dioxide to the atmosphere whilst sending treacle-like strands of red molten lava coursing down steep volcanic slopes. The second is the re-creation of granite underneath the continental margins, as the calcium, silicon and oxygen along with many other chemical beings re-configure themselves into the new rocky foundations of the continents, making up for granite lost from weathering the land surfaces. What perfect recycling of both granite and carbon dioxide! If only our industries could recycle their products so beautifully (Figure 22).

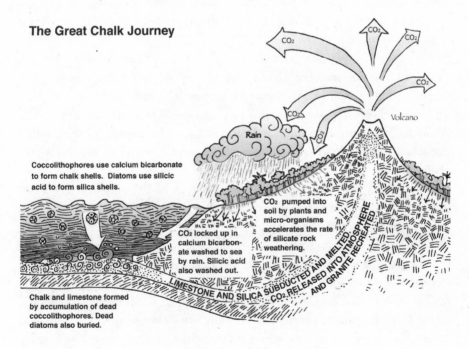

Figure 22: Carbon's longest journey.

Plate tectonics, driven as it is by the decay of radioactive materials deep in the Earth, seems to be totally independent of life. But nothing could be further from the truth, for without water there would be no plate tectonics, and without life there would be no water. Water molecules invade the crystalline matrix of the sea floor basalt as it moves away from the mid-oceanic ridges, softening it so much that by the time the basalt meets the edge of a continent it wants nothing more than to sink like so much semi-molten chocolate. But once in the basalt, some of the water molecules break up as their oxygen atoms feel an irresistible attraction for some key iron-bearing compounds, leaving hydrogen, the lightest of all chemical beings, free to escape to outer space. In time, so much hydrogen would be lost that Gaia would die of desiccation, a fate avoided thanks to the work of countless bacteria in the ocean sediments that capture energy by combining oxygen with the fleeing hydrogen, thereby re-creating the lost water and saving the planet. Once again we bend our heads in gratitude to the bacteria, the true rulers of the world. Furthermore, according to Don Anderson, the eminent American geologist, it is conceivable that plate tectonics could not happen without the calcium carbonate laid down by living beings at the edges of subduction zones, for the great weight of these sediments could be softening the underlying rocks, making them pliable enough to plunge into the deep Earth.

Volcanoes, hot springs and earthquakes are perhaps the most obvious manifestations of tectonic activity, but there are more unusual ways of encountering the immense churning energies deep beneath the crust of the Earth that keep her plates in constant motion. Once, as a member of a zoological expedition to the Caribbean island of Dominica, I hiked with a friend up to Trafalgar Falls, deep within the balmy forest. We clambered around many huge boulders that lay strewn along the lower reaches of the valley and, happily suffused with the blessings of the ample green forest and the tropical sunshine, we reached the peaceful solitude of the falls. The cascades were inviting, and as we bathed and wallowed beneath the luscious curtains of crystalline water we discovered a narrow geothermal torrent gushing hot water like none I had ever encountered. It was as if the experiences of its long journeys through the dark hot rocks far from the realms of air and light, of its long contact with semi-molten basalt and newly formed granite, permeated my skin like richly scented oil. This was a warmth charged with an unexpected communicative power. The cascading water informed my sensing body of the rocky underworld that lay beneath our feet, far below the threshold of our everyday awareness; a realm, in its own way, as animate and turbulent as the wind, rain and ocean of our daylight world.

This geothermal energy is vitally important, for Gaia would freeze without the return of carbon dioxide from the fusing and melting of calcium carbonate and silica beneath her moving rocky plates. But what if too much carbon dioxide returns to the atmosphere through volcanoes—wouldn't Gaia be seared to death by temperatures far too high for life to bear? This is indeed a mortal danger, but as we have seen, Gaia has remained wonderfully poised within habitable temperatures over much of her long life. How is this possible? It seems that the slow *tai chi* dance of negative feedback has kept the temperature right for life, a great dance that involves all the living beings, rocks and gases we have encountered so far, the dance of carbon's great chalk journey through Gaia.

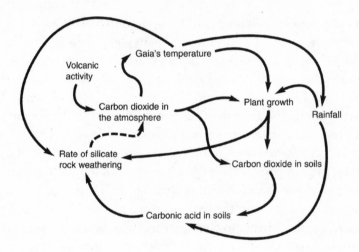

Figure 23: Carbon's longest journey depicted as a set of feedbacks that have kept Gaia's temperature within habitable bounds over geological time.

There are seven interlinked negative feedbacks involved in this great set of self-regulatory dances (Figure 23). Volcanoes have the important job of supplying the air with fresh legions of carbon dioxide molecules from the fusion of chalk and silica deep in the Earth. Note, however, that these great conical mountains of lava are free to behave as they wish because there are no couplings from the rest of the dance to control their tempestuous eruptions. Thus the entirety of the surface world must adapt and respond to the immense churnings of semi-molten rock deep in Earth's interior, of which volcanoes are one expression.

Let us journey around one of these feedbacks. First of all, look back at Figure 8 to remind yourself of what the two kinds of arrow (solid and dashed) mean. Now imagine that great volcanic eruptions have spewed vast amounts of carbon dioxide into the atmosphere. The whole of Gaia warms because of the increased greenhouse effect, and so more water evaporates from the oceans into the air, eventually condensing as rain-bearing clouds. Some of this rain falls on land where vegetation grows on granite or basalt. The life-giving rainwater percolates down through the soil to be absorbed by the plants, which grow better in the moister conditions. More rock is crumbled and ground up by roots, fungi and bacteria, which breathe large quantities of carbon dioxide onto the vastly increased surface area provided by a myriad of rocky fragments. This life-enhanced weathering of granite and basalt sucks carbon dioxide out of the atmosphere and sends it into the rivers as calcium bicarbonate ions where the carbon eventually finds its way into deposits of chalk and limestone on the sea bed. Gaia is cooler now, with less carbon dioxide in her atmosphere, so there is less rainfall, and in the drier world there is less life-assisted weathering of granite and basalt. The great dance comes full circle as volcanoes warm the planet through their return of carbon dioxide to the air.

Now let's follow another journey. Plants grow more vigorously in an atmosphere rich in carbon dioxide, an essential nutrient that they deftly capture through tiny pores on the undersides of their leaves. As the plants grow, their roots expand in the soil and weather more granite and basalt, thereby cooling the Earth. Now another journey. When carbon dioxide in the atmosphere increases, the higher temperatures stimulate the growth of plants, which increases the rate of granite and basalt weathering, thereby cooling the planet. I will leave you to trace out the remaining journeys on your own.

Now, after this fairly detailed exploration of the abstract terrain of cybernetics and Gaian self-regulation, we need to ask ourselves how we can use this approach to ground ourselves ever more deeply in our animate Earth. Indeed, there is a danger that cybernetics can leave us with a severe sense of disconnection from the world—that it can lead us, in the words of David Abram, to believe in an "entirely *flat* world seen from above, *a world without depth,* a nature that we are not a part of but that we look at from outside—like a God". How do we avoid this? I find it helpful to convert the feedback diagrams into *stories* by taking them outdoors, either literally or imaginatively, where I dissolve the flat-world pictures with their various arrows and components into an intuitive experience of the embedded relationships amongst the animate subjects that the story speaks of. If the story concerns the weathering of granite by rain and vegetation, I spend time with a granite boulder, sensing my way into the depths of its rocky silicate matrix, and then similarly with rain, with trees and moss, and with rivers and clouds. This practice opens up the senses and intuition, allowing one to move through the storied feedbacks in a kind of awakened dream state that yields a rich harvest of meaning, significance and belonging.

Breathing Chalk and Granite

If you can, find a small piece of granite and a small piece of chalk or limestone. If you can't find these stones, then imagine that you are holding one in each hand. The two stones represent all of the granite and all the chalk on our planet's surface. Make yourself comfortable, either indoors or out, and become aware of breathing slowly and naturally.

Focusing now only on your in-breaths, imagine that carbon dioxide is being drawn out of the atmosphere by the life-enhanced weathering of the granite in your hand. Imagine the tree roots, the microbes and the fungi crumbling the rock, surrounding it with water and carbon dioxide, which is locked up with calcium in a chalky solution of calcium bicarbonate washed down to the sea to be made into solid chalk by the coccolithophores. See their chalky shells sinking to the bottom of the ocean, and visualise vast deposits of chalk and limestone laid down on the sea bed—the very chalk or limestone that you are holding in your other hand. Become aware of this stone now.

Focusing now only on your out-breaths, see how the chalk sediments are pushed deep under the Earth when the movements of the Earth's tectonic plates collide them against a continent. Feel the carbon dioxide spewing out of volcanoes as the chalk melts in the intense temperatures and pressures deep down beneath the continent. As your out-breath stops, you become carbon dioxide coursing through the atmosphere, warming the Earth.

Feel how, in its melting, the chalk has contributed to the making of new granite. Connecting once again with the presence of the granite in your hand, repeat the cycle until it flows naturally and easily.

But what of the brightening sun? Why hasn't its increasingly generous gift of energy overwhelmed these great negative feedback loops, sending Gaia into an early heat death? The answer seems to involve life's creative expansion into novelty right from its first appearance on the planet about 3,500 million years ago. Since those early days when bacteria were the only life forms, until now, when Gaia teems with a huge variety of multicellular creatures and hosts a biodiversity greater than ever before, life has become better and better at weathering rocks and hence at drawing carbon dioxide out of the air into chalk and limestone. In Gaia's early days, under a cool sun, increasingly widespread and effective microbial films on granite surfaces carried out the weathering. In this early phase of her life, Gaia needed an atmosphere which could keep the surface warm in the face of a cool sun. With only bacteria present on a much smaller continental area, there was far less weathering of silicate rocks than is the case today. Later, about 2,500 million years ago, when the first nucleated cells spread over the larger continents, weathering increased. Even later, by about 400 million years ago, under a considerably brighter sun, there was a need for more effective weathering to remove carbon dioxide from the atmosphere. The newly evolved land plants provided this much-needed amplification of weathering because their hugely effective root systems were able to crumble and dissolve rocks with a speed and efficiency undreamt of by the microbial realm.

Thus, as Gaia has evolved, the diversification amongst her living beings has gone hand in hand with greater rock-weathering abilities and with more effective draw-down of carbon dioxide from the atmosphere into chalk and limestone. During this whole evolutionary dance the relationship between life, rocks, atmosphere and oceans has intensified and deepened like a good marriage, and Gaia has honed and augmented her skill at regulating her temperature. She has become more exquisitely responsive to both the sun's increasing brightness and to varying amounts of carbon dioxide released from volcanoes, much as a musician begins as a promising young player and matures into an accomplished virtuoso. As time has gone by, more and more players in the form of newly evolved species have added their voices to Gaia's symphony, so that today her orchestra burgeons as never before with diverse sounds from a rich variety of players and their instruments.

But Gaia has not always been able to regulate her temperature smoothly and easily in the face of the ever-brightening sun. In Figure 24, the horizontal axis shows time, running from about 600 million years ago till now, whilst the vertical axis shows Gaia's corresponding temperature. In the 'sun only' curve, we see how temperature would have increased as the sun brightened over time had there been a constant amount of carbon dioxide in the atmosphere. But this constancy is of course an illusion—it is nothing more than a mathematical 'static Earth' for the purposes of comparison.

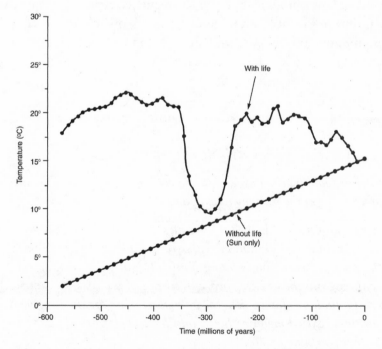

Figure 24: Global temperature with and without life
over the last 600 million years.

The upper curve reveals the actual temperature—it shows that there was a significant cooling around 300 million years ago, to the extent that a major glaciation gripped the planet. This event, which marks a major transition in Gaia's evolution into greater maturity, was caused by a massive increase in the biologically assisted weathering of granite and basalt as deep-rooted land plants took hold of the planet's land surfaces. It took Gaia about 100 million years to recover from this self-imposed perturbation. By around 200 million years ago there was much less granite and basalt available for weathering, and so carbon dioxide once again began to accumulate in the atmosphere. Amazingly, temperatures before and after the cooling were remarkably similar. Before the event some 600 million years ago, under a dimmer sun, higher levels of carbon dioxide kept temperatures well within habitable levels. Since her recovery from the cooling, Gaia has dealt with the sun's increased brightness by safely tucking carbon away in carbonate sediments at the bottom of the sea and by burying the bodies of dead carbon-rich photosynthetic beings.

We have been working for some time now with a cognitive way of understanding how our animate Earth has regulated her temperature over geological time in the face of a brightening sun. Now it is time to engage in a more intuitive way of entering into this knowledge, by means of an extended experience of the long-term carbon cycle.

A Carbon Journey

Go to your Gaia place, and take in the colours and sounds that surround you. Begin the journey when you are calm and relaxed.

You are a carbon atom locked up in limestone rock at the bottom of the sea. For the last 300 million years you have experienced nothing but the cold pressure and seeming immobility of solid rock around you. Nothing but the ancient, peaceful silence of the rock. No movement, no sound, no change, just the immense repose that has engulfed you for time beyond memory. Settle into the feeling of immense tranquility that surrounds you.

You experience a slow sinking feeling as the sea floor that carries you from below slowly approaches the edge of a continent, dragging you downwards into the depths of the Earth.

Feel the temperature and pressure increasing as you sink down into the dark depths. It's so hot now that the limestone you are part of begins to melt, merging with the silica that surrounds you.

Feel the intense heat and pressure. The chemical bonds holding you to your calcium atom vibrate with an agonising intensity. You are shaken about like a pea in a box. The shaking is so extreme now that you realise that you are being liberated from the limestone.

You feel awake and excited. You tingle with anticipation.

Feeling incomplete, you quickly bond with two passing oxygen atoms newly released from the glassy shell of a melting diatom.

Savour the passionate embraces of your two oxygen lovers. The three of you are now part of a new emergent being—a molecule of carbon dioxide.

Now the red hot, molten rock around you flows quickly upwards. You travel faster and faster through a great wide gash in Gaia's crust, almost deafened by the mad sound of surging gas and rock. The sound intensifies, and with it the slow, red-hot turbulence that carries you higher and higher, closer and closer to a new life that you know will soon open up for you.

Pressure is building behind you. You are moving very fast. The pressure, the sound, and the heat are immense and powerful.

You are right in the very heart of a massive volcanic eruption. A bursting release of energy propels you high into the air along with vast amounts of smoke, ash and red hot lava.

You look down at the smoking volcano far below. Already you are high up in the atmosphere, buffeted by the intense updraft that has brought you this far so quickly.

All day long the sun has been beating down. You drift in the air, warming it as heat from Gaia's surface bounces off the bonds that hold you to your two oxygen lovers.

Great air currents carry you northwards, and for weeks you soar high above the seas and forests. You enjoy the delicious freedom of travel, wondering at the amazing views of mountains, forests and oceans, bonded still to your oxygen lovers. How the planet has changed since you were last here, some 300 million years ago, when the first four-limbed creatures crawled upon the Earth.

At last a great gust of air carries you down over a modern-day granite outcrop. You swirl closer and closer to the ground, and sweep past a succulent bush, alive with luscious yellow flowers. You brush against a pore on the underside of a leaf, and are caught in a tiny in-breath that sucks you into the translucent green interior of the leaf. By depriving the air of a carbon dioxide molecule, your leaving cools the Earth.

Giant molecular beings surround you and take you to a great green chamber deep in the cell—the chloroplast. A blinding flash of sunlight shakes you to your core. You are joined to other carbon and hydrogen atoms. The newly forged chemical bonds that bind you to your fellow atoms hold the sun's energy. You are now part of a sugar molecule, beginning a new journey through peaceful green sap.

You feel a tugging downwards, towards the plant's roots. Slowly you move through wide tubes, ever downwards, pulled by a ceaseless but subtle flow that takes many other molecular beings along with you.

You reach the very tip of a growing root hair and pass into a growing cell. All around you there is frenetic activity as new root cells are made, pushing the root ever further in its incessant search for nutrients. Your root finds a crack in the solid granite beneath the soil. It breaks into the crack, splitting the rock as it swells.

Once again oxygen lovers embrace you, tearing you apart, releasing the solar energy you have held for so long. Now, once again, you belong to a carbon dioxide molecule, breathed out by the root into the surrounding soil.

There has been much rain, and the soil is waterlogged. You feel an irresistible attraction for a passing water molecule, and together you become carbonic acid, which instantly releases a single, tiny hydrogen ion, the smallest being in the chemical world. The carbonic acid molecules around you release a vast horde of hydrogen ions into the soil.

The hydrogen ions dissolve the granite, releasing calcium and silicon atoms, long incarcerated in the rock. You feel irresistibly attracted to one of these calcium atoms as it drifts near you. You bond with it to become liquid chalk.

You are pulled downwards by the flow of ground water. Now your great river journey begins. You hear a rushing, gushing sound, and suddenly you enter the flow of the river as it tumbles over great boulders and waterfalls on its way to the sea.

In the cool surface waters of the sea you are sucked into the embrace of a marine alga protected by tiny wheels of solid chalk. You soon become part of one of these wheels.

The alga lives its short life and dies. You slowly sink towards the ocean deeps, cherishing the memories of your brief journey in the air above.

Sinking, sinking into the depths, you leave behind the upper sunlit reaches of the sea and eventually settle into the chalky sediments in the darkness at the bottom of the ocean.

Slowly you feel the weight of new chalky deposits accumulating above you, and slowly, ever so slowly, the pressure packs you closer and closer together into limestone. The great journey is complete, and now you must wait for another 300 million years before you once again explode into the atmosphere through a volcano.

Carbon in the Short Term

People often react to the story of carbon's great chalk journey by asking whether the carbon dioxide we are adding to the atmosphere through our manic burning of fossil fuels could be rapidly removed in this way. Sadly, the process is too slow to make any difference over the next century or so. It may take a carbon atom emitted into the atmosphere in the diesel exhaust of the train that I'm riding on today about half a million to a million years to find itself in chalk or limestone at the bottom of the ocean—far too long to make any difference in preventing climate change.

So what about carbon dioxide's journeys in the much shorter term? Carbon travels through Gaia through a set of nested loops, each operating at its own particular pace. A carbon atom can get swept into any of these journeys, depending on where it happens to be at crucial moments, and it is very likely that virtually all the carbon atoms on the planet have experienced what it is like to pass through each of them. Amongst the shortest of carbon journeys are those that take as little as a year to complete. The global effect of this cycle can be seen in the now famous data collected from the Mauna Loa observatory in Hawaii (Figure 25)—the first warning of our serious impact on Gaia's climate.

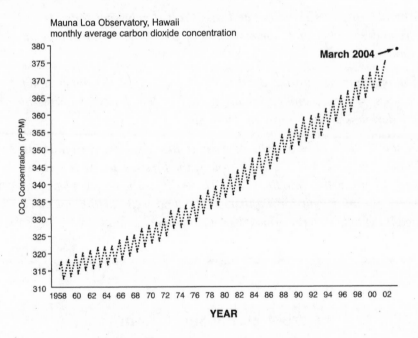

Figure 25: The data from Mauna Loa.

High up on the Mauna Loa volcano, one finds air stirred together from all over the northern hemisphere, with very little contamination from local sources. Scientists analysing this air found a regular oscillation, once again a bit like the 'breathing' we saw in the Vostok ice core data (Figure 5), but this time with a yearly rhythm. What drives this cycle? If you are a carbon dioxide molecule in the air somewhere in the high latitudes, it is possible that you'll be sucked up into a land plant through one of the minute breathing pores, or stomata, on the underside of a leaf during the spring or summer when there's plenty of moisture, light and warmth to fuel photosynthesis. Soon you become part of a sugar molecule deep within the leaf. Then, in the autumn, your leaf is shed and tumbles helter-skelter towards the ground. There a soil microbe absorbs the sugar molecule in which you find yourself. The tiny microbe uses oxygen to extract the energy in the chemical bonds that bind you to your fellow carbon atoms, and you are once again linked to two oxygen atoms and expelled into air through the microbe's semi-permeable membrane. You've been cycled out of the atmosphere and back again in a molecule of carbon dioxide within a single year.

In summer, photosynthesis takes out much more carbon dioxide than is released from the soil by microbes breaking down dead leaves and other organic material, producing the troughs in the Mauna Loa graph. In autumn and winter, photosynthesis shuts down in the high latitudes as plants go into a deep dormancy. But the soil microbes continue their work of breaking down organic carbon despite the low temperatures, and so carbon dioxide wafts out of the soil and into the air, showing up as the peaks in the graph. This regular, yearly breathing is more marked in the northern latitudes because there is more land mass on which the vegetation can grow—in the southern hemisphere there is far less land and so the breathing is weaker. The upward trend in carbon dioxide concentration is primarily due to our emissions of this gas.

Although the soil is a great reservoir of carbon fixed by photosynthesis, it seems that what happens in the soil could, paradoxically, help to warm Gaia if we continue to emit carbon into the atmosphere at very high rates. Recent models from the Hadley Centre in the UK show that the activity of the soil microbes increases dramatically when atmospheric carbon dioxide is at about 550 parts per million—a situation we will most likely reach within a few decades if emissions continue unabated. Higher levels of carbon dioxide in the air increase temperatures in the soil, which in turn stimulates the growth and activity of soil microbes. This leads to a massive breakdown of soil organic carbon, and hence to an outflow of carbon to the atmosphere that outstrips the inflow of carbon into the green world from photosynthesis. Here we see an important feature of non-linear systems such as our planet—what was once a negative feedback can easily become a dangerous positive feedback if the system is forced beyond a critical tipping point.

So far, we've looked at how carbon moves in and out of soils and land plants, but carbon's journeys through the oceans are also hugely important. The ocean absorbs around 30% of the carbon dioxide currently emitted into the atmosphere by humans. Scientists talk about three major 'pumps' by means of which the oceans remove carbon dioxide from the air—the solubility pump, the biological pump and the physical pump. The solubility pump doesn't directly involve life—it works simply because carbon dioxide dissolves in sea water. Waves break on the ocean surface, and as they ripple and swirl, carbon dioxide from the atmosphere is folded into the ocean much as a baker kneading bread wraps air from the kitchen into her creation. Because the gas molecules have less energy available for leaping back into the air if the water is cold, more carbon dioxide enters the ocean through this route in the high latitudes. In fact, carbon dioxide has a great talent for dissolving itself in water—it is twice as soluble as oxygen at 20^0C—a property that makes a huge difference to Gaia's temperature. The ocean surface is like a membrane across which carbon moves between air and sea. If carbon dioxide is added to the air, there is a flow into the ocean via the solubility pump; if some of the gas is removed from the air, it flows out from the ocean, until, in both cases, equilibrium is reached.

The biological pump is itself divided in two—the 'biological organic pump' and the 'biological carbonate pump', already known to us as Gaia's great chalk journey. Let us now take up the story of a carbon atom as it experiences a journey through the biological organic pump. After being exhaled by a microbe in the soil, our carbon atom finds itself wafting in the air over the ocean, linked with two oxygen brothers in a molecule of carbon dioxide. Suddenly a wave catches it, and our carbon dioxide molecule is dissolved in the clear ocean water. A passing alga, photosynthesising in the bright sun, absorbs it, and soon the dextrous metabolism of the cell links our carbon atom to a multitude of other atoms in the long chain of a vast sugar molecule. Now our carbon atom floats in the lovely translucent greenness of the alga's body, waiting for oxygen to break open the sugar molecule, releasing its embodied solar energy. But before this happens, the alga is devoured by a tiny amoeba-like predator known as a radiolarian, which extends its engulfing pseudopods through tiny gaps in its exquisite silica shell. Many other predators roam the surface waters—small fish, tiny free-floating shrimp-like carnivores (the copepods), and jellyfish raking the water with their long stinging tentacles for their sugar-rich prey. The radiolarian is itself eaten by a copepod, which digests much of the carbon that was in our alga's body, but the particular sugar mole-cule in which our carbon atom happens to find itself evades this fate, and is excreted into the water in a faecal pellet along with the remains of several other algae. As the heavy faecal pellet sinks down towards the depths of the ocean, translucent green is slowly replaced by dark green, then by deep blue and eventually by pitch blackness as our carbon atom reaches the abyss. During this long downwards journey a host of creatures devour most of the faecal pellets, releasing carbon dioxide into the sea. But the faecal pellet that has transported our carbon atom into the depths is one of the few that survives, and it settles at last on a layer of soft mud on the seabed. Even here, there are creatures that eat and digest faecal pellets, but our carbon atom avoids this fate, and is gradually covered in silt and sand washed in by a river draining the nearby continent. It is now one of the tiny proportion of carbon atoms to have reached the deep sediments from the surface ocean via the biological organic pump. Only 1% of the organic matter in the sinking mass of faecal pellets arrives on the sea floor, and of this only 0.1% is buried in the sediments.

Phosphorus and nitrogen from the surface waters are also transported to the ocean depths by the biological pump, greatly affecting the chemistry and ecology of the surface ocean, and, as we shall later discover, the oxygen content of the atmosphere. Organic carbon from the land in the form of tiny fragments of undigested leaf and wood also makes its way into the ocean sediments, transported by the great rivers of the world. About 25% of all the organic carbon finding its way into the deep ocean stays there for 100 to 1,000 years, but the story of our carbon atom is different. It experiences gradually increasing pressure as more and more sediment accumulates above it, until after millions of years it has become part of a sedimentary rock deep below the sediment surface. Here it will stay, locked up in a rigid casing of compacted, chemically transformed sediment, for perhaps 200 million years. Strangely, most of the organic carbon on our planet resides not in living cells, but as the remains of once-living beings scattered like confetti in the world's sedimentary rocks. This 'sedimentary organic carbon reservoir' is huge because the outflow from it is very small, and because there has been enough time for vast amounts of carbon to accumulate in it despite a very small inflow.

Eventually our carbon atom feels the rock around it being slowly uplifted as the oceanic plate which has carried it along collides with a continental land mass. As the great forces crumple the rock against the continent, our carbon atom becomes part of a magnificent mountain chain that condenses clouds out of the moist air on its windward flanks. The rain tumbles out of the sky and weathers the rock, until one day a small flake falls off a sheer face and our carbon atom feels the enlivening whirling of the atmosphere around it for the first time in hundreds of millions of years. An oxygen molecule, in its hunger for electrons, burns up the very sugar molecule that has held our carbon atom in a seemingly eternal embrace throughout this long journey, and it returns once again to the atmosphere in a newly minted molecule of carbon dioxide. It feels wonderful to be in the atmosphere after such a long absence, and for ten years our carbon atom roams the world, evading capture by photosynthesis. Travelling the world's air, it sees the great oceans, and the great forests, until one day a pore on the underside of a leaf on a great Amazonian tree captures it. Now, for a brief while, it will experience the qualities of leaf and wood before continuing on its never-ending explorations within the body of the Earth.

Great downwellings of carbon-rich ocean water in the high latitudes constitute the third pump—the physical pump (Figure 26). As the sun beats down on cloudless tropical oceans, great tongues of warm waters are carried into the high latitudes by strong currents such as the Gulf Stream. Much of the warmth leaves the water as it travels away from the tropics— some is transferred directly to the surrounding air and some is carried off by water evaporating from the sea's surface. Either way, the escaping warmth leaves behind dense, cold, salty water. By the time the Gulf Stream reaches Greenland it is so much denser than the surrounding water that it sinks and plunges down in two huge underwater surges to the bottom of the ocean. A similar sinking takes place around Antarctica. Significant amounts of carbon—as carbon dioxide dissolved directly from the air and as the carbon-rich corpses of a whole legion of marine organisms—are dragged down into the ocean depths with the downwelling waters. This deep, cold, carbon-rich bottom water streams along the sea bed for thousands of kilometres until it branches in the abyss of the south-west Atlantic, surfacing about a thousand years later in the Indian ocean and in the central north Pacific, releasing its hidden cargo of carbon dioxide to the air as the waters warm once more under the hot tropical sun. The water travels through the tropical oceans, pushed along mainly by the power of the wind, warming as it goes, until it completes its global journey as the great Gulf Stream that brings so much warmth to the north-east

Atlantic. If our planet has an equivalent of flowing blood, this must be it, for this vast global flow of water ably distributes dissolved gases, warmth from the sun and vital nutrients around Gaia's great spherical living body.

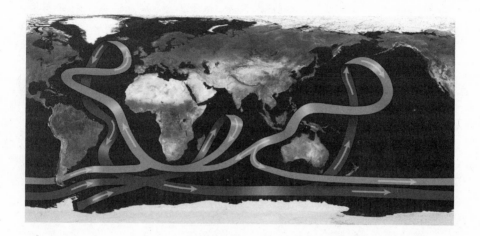

Figure 26: The global thermohaline circulation in the world's oceans.
(© *Science Photo Library*)

But this global circulation of ocean water is a delicate thing, highly vulnerable to changes in the density of sea water in the downwelling regions around Greenland. Towards the end of the last ice age, as the world warmed, huge amounts of fresh water from the melting of the North American ice caps entered the North Atlantic, freshening the warm salty waters of the Gulf Stream to such an extent that their sinking was prevented. As a result, the downwelling weakened or shut down around 12,900 years ago, plunging Europe and the entire North Atlantic into a dramatic cooling event known as the Younger Dryas, that delayed the beginning of the current warm interglacial by some 1,400 years. The effects of this cooling were felt far beyond Europe. Even more dramatic was the end of this period of intense cold, in which an abrupt warming of around 7⁰C took place in a decade or so. Herein lies another warning for contemporary society—it seems likely that our war on nature is set to trigger similarly abrupt, catastrophic changes to our climate—with very little warning.

Chapter 6

Life, Clouds and Gaia

Sulphur and Albedo

So far we have seen how life is profoundly involved in regulating Gaia's temperature through its impact on the gaseous chemical beings that inhabit her atmosphere. But this is not the only way that life has contributed to Gaia's emergent ability to maintain a habitable planet—it also has the power to alter her overall shade, or albedo.

Living beings do this in two major ways. One is by releasing chemicals that seed vast banks of dense white clouds that reflect the sun's energy back to space before it has had a chance to heat up the Earth's surface. Another is by covering vast areas of land with dark or light vegetation that respectively absorb or reflect the sun's rays, thereby warming or cooling the Earth. Until fairly recently, scientists believed that a given ecological community simply responded to whatever environmental conditions had been bestowed upon it by the great 'non-living' forces of the planet, such as plate tectonics, so it came as quite a shock to them to discover that this is only partially true; for in fact whole ecological communities, both in the ocean and on land, affect their local temperatures simply by altering their surface albedo, much as the inhabitants of houses in sun-drenched deserts paint their houses white to keep them pleasantly cool inside.

This unexpected ability of life to alter planetary albedo has to do with the global sulphur cycle. Sulphur is essential for life—without it the amino acids that build proteins cannot be made—but it is very scarce in the soil, and is hence a limiting nutrient for plants. In the ocean, however, we find plenty of sulphur, brought there by the rivers that garner it from the weathering of rocks exposed to the air. There might at first seem to be a contradiction here: how is it possible that sulphur is scarce in the soil if its initial source is the weathering of sulphur-rich rocks? The answer is that most of the sulphur is whisked away by rivers to the sea before life on land can get its roots into it.

Gaia faces the critical problem of transporting the sulphur from the ocean where it is abundant to the land where it is scarce; for without this vital transfer, terrestrial life would be impossible. James Lovelock predicted that organisms in the ocean had to be involved in this process by producing a sulphur-carrying gas that would blow in on ocean winds to fertilise the land. But what could the gas be? An obvious candidate is hydrogen sulphide, the gas beloved by children in their early dabblings with chemistry because of its pungent odour of rotten eggs. Hydrogen sulphide, however, can't do the trick. Firstly, not enough of it is produced, and secondly it is quickly and passionately dismantled by hydroxyl ions, the aerial children of oxygen, which find it irresistibly attractive.

Next time you go down to the sea, take a deep breath. The delicious tangy, uplifting aroma that greets your senses is the gas dimethyl sulphide, produced by marine algae. The tanginess comes from sulphur in the gas, which gives it a slightly acidic bouquet. During a long sea voyage, Lovelock found plenty of DMS in the air above the ocean, and calculations by his colleagues later confirmed that the sea produces more than enough DMS to close the sulphur cycle.

But DMS is important not just as the carrier gas for sulphur; it also plays a vital role in seeding planet-cooling clouds. Seen from space, the Earth looks like a beautiful blue marble streaked with swirling pearly white mountains of water vapour. These are Gaia's clouds, the silent captains of the sky. At any given moment, much of her surface is covered by them, and some, generally the low ones such as marine stratus, cool the Earth by reflecting the sun's light to space from their dense white upper surfaces, whilst others, the high fliers such as cirrus, warm the Earth by delaying the exodus of heat radiated from the surface.

Everyone knows that clouds appear when water vapour from the ocean and the land condenses in the air above us, but it takes more than just water vapour to make a cloud. The water molecules swarming about in water vapour would like nothing better than to clump up very close together to make a cloud, but they can't do this by themselves—they must first condense on small particles in the air known as cloud condensation nuclei (CCN). Particles of dust blown in from the land do this job well, as can salt spray sucked up by winds from the ocean surface, but both are nowhere near common enough to account for the abundant swirling whiteness that cloaks our planet.

For a long time no one knew what the mysterious cloud-seeding particles were, or where they might be coming from, but Lovelock had a strong intuition that organisms must be centrally involved in producing them. Eventually, he helped to make the discovery that marine algae such as *Emiliania huxleyii*—the very same beings who contribute so much to regulate Gaia's temperature by precipitating chalk—are major players in this process.

It turns out that *Emiliania*, and indeed many other algae such as seaweeds, emit DMS which finds its way into the atmosphere where it attracts the ardent attentions of oxygen. Remember how oxygen, the passionate Italian of the chemical world, finds completion by seducing electrons away from other atoms or molecules in order to complete its own outer electron orbit. The gas DMS is a great target for oxygen's ardour, and when it has sucked the electrons it needs from this larger molecular being, what remains, amongst other things, are molecules of sulphate aerosol floating free in the air above the ocean. These molecules have many special qualities, but one of critical importance for Earth's climate is that water vapour finds them irresistibly attractive, preferring nothing better than to condense around them in dense schools and shoals like so many fish swarming around bread crumbs thrown into a lake.

The massive condensation triggered by molecules of sulphate aerosol derived from DMS makes clouds that cool the Earth, because their dense white upper surfaces—familiar to anyone who has travelled by air—reflect solar energy back into space. But can this astonishing relationship between algae and clouds help to regulate the Earth's temperature by means of a negative feedback in which warmer oceans stimulate algal growth, resulting in more DMS, more sulphate aerosol, more cloud condensation nuclei and hence more planet-cooling clouds? This feedback, proposed by Lovelock and his colleagues Robert Charlson, Stephen Warren and Meinrat Andreae, in its simplest form looks like this (Figure 27):

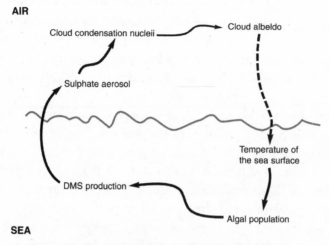

Figure 27: The proposed negative feedback involving marine algae and the clouds they seed.

We now know that proposal is far too simplistic, and that the relationship probably switches from positive to negative feedback depending on circumstances. The precise details are not yet clear, but what is certain is that algae in the ocean are crucially important in generating clouds and therefore in contributing to global climate. But a further puzzle remains. Why should marine algae bother to expend large amounts of hard-won energy in making DMS gas? Evolutionary biologists have pointed out that the algae cannot be doing this for altruistic reasons—they must be gaining a direct and immediate benefit for themselves. One answer to this question lies in the chemical nature of the precursor to DMS, a much more complex molecule, dimethylsulphoniopropionate, known as DMSP.

We are inside a coccolithophore cell floating in the sunlit waters of the ocean surface. All around us, a host of complex molecules and a myriad of sodium, calcium, chloride and other ions engage in the highly ordered but apparently random dance of metabolism. There is one very simple molecule that is essential for keeping this entire metabolic hurly-burly going—water. Without it, everything would grind to a halt as the positive and negative charges that lie scattered amongst the denizens of the intracellular molecular world attract each other into a messy, gooey stasis. But water molecules, each a combination of two negatively charged oxygen atoms and a positively charged hydrogen atom, gather around and shield the positive and negative charges of their molecular neighbours, preventing the gumming up and allowing the metabolism of the cell to flow smoothly.

But, strangely enough, these same attractions between water molecules and positive and negative ions in the surrounding sea water outside the cell threaten the very life of our alga because of a peculiar force known to biologists as *osmosis*. Let's swim close to the alga's cell membrane that encloses the entire cell in a sinuous, filmy, balloon-like embrace. Notice that the membrane is riddled with tiny pores. If you watch carefully, you'll notice how the random motions of the water molecules result in many of them leaving our algal cell through the pores to enter the sea, whilst other water molecules find their way into the cell from the surrounding sea water, also via the pores. If we were able to paint the water molecules inside our alga with a red dye, and those in the sea water immediately surrounding us with a blue dye, we would quickly notice that more red water molecules end up outside the cell than are replaced by blue ones coming in from the outside. In other words, our cell loses more water than it gains from the surrounding ocean.

Why should this be? Unlike most molecules and ions, water can move freely in either direction across our alga's semi-permeable cell membrane. Let's hitch a ride on a passing water molecule as it travels out through a pore and into the surrounding sea water. We look back at our cell hovering before us like a giant planet in the vast spaciousness of the sea, where there are many more ions than inside our cell. Like so many other water molecules in the ocean, ours is attracted to these more abundant ions to such an extent that fewer water molecules are free to pass into our alga through its cell membrane. But inside the cell there are fewer ions and charged molecules to attract the water molecules, and so more of them leave the cell as their random movements tumble them out into the sea through the pores in the cell membrane. The result is that the entire cell is in danger of dying, as the unshielded ions gum up the cell's metabolism.

Our alga can prevent water loss to a certain extent by using energy to pump water back into the cell from the sea, but it can't bring in enough water to prevent an uncomfortable situation from developing. And now at last we come to the role of DMSP—a relatively long molecule with a positive charge in the vicinity of the sulphur atom at one end, and a negative charge at its opposite end. These two charges attract free ions that might otherwise wreak havoc the cell. The situation looks rather like this (Figure 28):

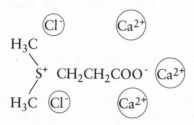

Figure 28. The DMSP molecule, with positively and
negatively charged ions (circled) clustering around it.

When the algal cell dies, or is eaten by predators, the DMSP rapidly
degrades into DMS gas, so in this scenario DMS is no more than a mere
by-product.

It turns out that there might be another benefit of making DMSP—
protection from predators. The algae make an enzyme which breaks down
DMSP into DMS and acrylic acid, a rather acerbic, foul-tasting molecule
which is very good at deterring small fish and planktic unicellular carnivores, such as the silica-secreting radiolarians that love to eat the algae;
but here, once again, DMS is again no more than a mere by-product.

Could there be a direct benefit from producing DMS? After food and
sex, dispersal is of major importance for living things, including our tiny
oceanic algae. Ocean currents have stirred up sediments from the bottom
of the ocean, and all sorts of algae have taken advantage of the nutrient
bonanza coming up from the depths, until the nutrients are almost
depleted. Most DMS-emitting algae, such as our coccolithophores, bloom
in vast numbers when the nutrients have almost disappeared, perhaps
because they experience less competition from other species in nutrient
poor waters. But it isn't too long before nutrient levels become so low that
the spectre of starvation looms large. From the perspective of a single
coccolithophore cell, the situation seems hopeless. The cell hasn't got the
physical capability to swim to a new nutrient-rich patch of ocean, and if
it's summer time and a lid of warm water has formed on the ocean surface,
there is no possibility that currents will bring up fresh nutrients from the
sediments below. But the algae seem to have developed a cunning way of
averting disaster.

We are back inside our coccolithophore, which has sensed the fact that
nutrient levels in the surrounding sea water are becoming dangerously

low. Suddenly we see huge numbers of DMSP-digesting enzymes appearing as if from nowhere to begin the conversion *en masse* of DMSP to DMS gas and acrylic acid. The DMS gas rapidly leaves the cell through the cell membrane, journeying into the air via the sea, and the acrylic acid helps to deter passing predatory beings. As billions of tiny algal cells respond in this way to the nutrient crisis, the air above the bloom receives a great pulse of DMS that triggers the condensation of massive, dense white clouds above the ocean. As the clouds form, a huge quantity of energy is released as heat that makes the clouds rise. The newly forming clouds literally hoist themselves aloft, sucking air in behind them, much as air is drawn into a bicycle pump when the plunger is pulled outwards. The coccolithophores begin to experience wave action as the updrafts of air stir the surface of the ocean, and if our cell happens to be close to the water surface during a strong upward gust, it stands a good chance of being sucked up into the air and into a rapidly ascending cloud. What an experience it must be for a tiny sub-microscopic speck of marine life, which until now has known only a watery ocean existence, to be so suddenly wrenched up into a cloud that could carry it for hundreds of kilometres before washing it out in rain into a new region of the ocean where, with luck, there is an abundance of nutrients.

The most prolific DMS-producing algae are by far the smallest, perhaps because their diminutive size increases their chances of making it into a cloud. These tiniest algae also make foams and slimes that make it even easier to get airborne, and many of them have a red pigment that protects them from the abundant ultraviolet light of the high altitudes. One of the surprising properties of DMSP is that it is an anti-freeze, which works to the alga's advantage, for as the cloud rises, it reaches regions of low temperature where its water freezes out, only to fall earthwards as rain or snow. Thus DMSP may help the algae to survive not only the saltiness of the ocean, but also the freezing temperatures in clouds that the algae themselves have seeded.

Riding the Clouds

Find a comfortable place outdoors to watch clouds. Relax, and then look carefully at their shapes and textures. Watch their movements in the sky, and then imagine myriads of microbes riding in the clouds, triggering rain that washes them out into pastures new. Imagine what it would be like to be a microbe swept into the cloud by an updraft of air.

Now shrink yourself down smaller and smaller until you become a coccolithophore surrounded by billions of other coccolithophores in a vast bloom floating on the surface of the North Atlantic just off the coast of England. Feel the coolness of the sea water as it caresses your chalky shell. Notice the tangy taste of the sea, and how it gently rocks you from side to side.

It is the end of spring, and the increasing warmth of the sun has created a layer of warm water on the surface of the sea that prevents currents from bringing you nutrients from the depths below. You feel an aching hunger as the nutrients gradually run out.

The whole bloom, with you included, begins to sweat billions of cloud-seeding dimethyl sulphide molecules into the air above you. Winds stir the surface of the waters as the dimethyl sulphide molecules trigger the formation of dense clouds above you.

Suddenly you feel yourself caught in a wave that reaches skywards, and you are pulled into an updraft of air. Up and up you go, buffeted by the winds, until you reach the very centre of a large marine stratus cloud, which slowly moves northwards towards the pole.

For days you travel in your cloud, sensing how it helps to keep the Earth cool by reflecting the sun's energy back to space from its dense white upper surface. You look out beyond your cloud and see a whole host of similar clouds stretching off in all directions. You feel grateful

*for anti-freeze molecules deep inside your unicellular body that pro-
tect you from the intense cold.*

*Eventually, after many days, the water vapour around you begins to
condense into rain, and you find yourself tumbling seawards out of
the sky in a rain drop. You meet the sea with a gentle splash, happy
once more to be in familiar surroundings. You have been lucky, for
the patch of sea you have landed in is rich in nutrients brought up by
currents from below. Your long fast is over, and you feast to your
heart's content on the nutritious sea water.*

It is not just algae floating in the ocean surface that emit DMS; it has
recently been discovered that coral reefs do it too. Corals are sedentary,
colonial animals related to sea anemones and jellyfish that secrete the
chalky armoured spires that make coral reefs such architecturally
complex, achingly beautiful places. Each coral animal is in fact a small sea
anemone, which feeds by capturing tiny planktic creatures on tentacles
peppered with stinging cells. But closely packed inside cells of the coral
animals are symbiotic algae, called dinoflagellates, that provide their hosts
with the sugary blessings of photosynthesis, in return for protection and
nutrients. Experiments have shown that these symbiotic beings release
DMS when they experience stress from increasing temperatures or excessive ultraviolet light. Prodigious amounts of this premier cloud seeding
chemical have been detected over Australia's Great Barrier Reef, raising
the exciting possibility that corals may seed clouds that cool their immediate surroundings when the sea water becomes too hot for comfort, or
when ultraviolet radiation from the sun is too intense. If so, we would
have living proof that coral reefs could be operating a powerful negative
feedback that would counteract the dangerous effects of global warming.

Clouds teem with life: they are full of microbes of all kinds that use them
as gigantic dispersal 'buses'. Some of these microbes are so comfortable in
their lofty homes that they have been found happily reproducing thousands
of feet above the surface. But it's not just marine microbes that seed clouds;
life on land does it too. Let's zoom down into the spiky bracts of a bromeliad,
a kind of aerial pineapple which lives on the branches of trees in the rain-

forests of south and central America. Bromeliads are not parasites; they don't use their roots to suck life-giving sap from their host, but use them instead to girdle a branch in a knotty tangle of tightly gripping lassos which keeps the bromeliad firmly in place. Bromeliads are merely epiphytes: they use the tree only for support, and must therefore get all their moisture and nutrients from rainwater. Bromeliads are perfectly shaped for water collection. Their thick, fleshy leaves act like spear-shaped gutters which together make a set of nested cup-like receptacles at the base of the plant where water collects in miniature ponds inhabited by complex ecological communities, each differing slightly from bract to bract and bromeliad to bromeliad. The water in these bromeliad ponds swarms with microbes, many of them algae that provide photosynthetic food for the whole community. Strangely enough, the algae live in a medium somewhat resembling sea water, and face an osmotic challenge similar to that encountered by their marine relatives. The algae appear to solve their osmotic problems by making DMSP. Thus large quantities of DMS waft from bromeliads into the air above the forest, where oxygen converts it into the cloud-seeding sulphate aerosols. It isn't just the bromeliad microbes that seed clouds; the trees do it too, by emitting not DMS but complex organic molecular beings called turpenes and isoprenes that oxygen dismantles into cloud condensation nuclei of another sort.

The entire forest not only seeds its own clouds, but also recycles the very water that makes the clouds by capturing rainfall with its roots before sending it out into the air through countless billions of microscopic pores on the undersides of its leaves. Most of the Amazon's rainfall comes from outside its basin, mainly on moist, warm winds from the Atlantic, but 25% of the Amazon's water is recycled as rain by the forest itself. This is far more than the Mississippi basin, which can only manage to recycle 10% of its water. The Amazon is a very efficient water-harvesting organism: of the water that falls on it, about half is returned to the air via leaves, whilst the rest feeds its extensive network of rivers and oxbow lakes. The clouds the forest seeds are planet-cooling—they are the dense white kind that bounce the sun's rays right back to space. When clouds condense over the Amazon forest, a vast amount of energy is released—about 40 times the entire energy use by humanity in a year. Much of this energy is shunted around the globe in great waves of air to affect the climates of far-flung regions such as North America, South Africa, South-east Asia and parts of Europe.

Cutting the forest down is a catastrophe, not only for the millions of species living in them, but also for the world's climate. The latest models from the Hadley Centre in Britain predict that the Amazon could soon collapse into tropical savannah, even without the cutting down of a single tree. Why? The Amazon seems to do best during an ice age, when the high latitudes are covered in ice but the tropics are deliciously balmy and just right for the cloud-seeding forests. In between ice ages, during the warm interglacial periods, the whole world warms and the forest can just about manage to keep its temperature within tolerable limits by frantically producing a thick sunshade of clouds. But in our times, with more carbon dioxide in the air and higher temperatures than have been seen on Earth for over 400,000 years, the forest can no longer cope. Soon it could lose more water through evaporation than it can capture by triggering rainfall through cloud formation, and could gradually dry out until a critical threshold is reached. Then it would die back exponentially fast, just like white daisies when the sun pushes them beyond their ability to regulate Daisyworld's temperature. The savannah which replaces the rainforest seeds fewer clouds, so the entire Earth would warm as a massive temperature gradient builds up between the tropics and the high latitudes, creating severe storms and hurricanes which wreak havoc over Earth's surface like demented genies bent on revenge for the foolishness of our kind.

Meeting at the Heart of the World

Every denizen of the rainforest contributes to the production of clouds by the forest as a whole, no matter how far removed from trees, bromeliad-dwelling algae and clouds it might appear to be. Indeed, the whole planetary web of life ensures that the climate of the rainforest remains healthy enough for trees to send their gifts of water and cloud-seeding chemical beings into the air.

I was once vouchsafed a special experience of one of these beings. I had joined some friends on a six-week expedition to the remote Roraima region of Venezuela adjoining the Brazilian Amazon. Conan Doyle had set his book 'The Lost World' amongst the huge flat-topped mountains (tepuis) of this region. We had travelled for about two weeks in a dugout canoe powered by a small outboard motor up the Cuyuni river, camping on river beaches surrounded by lush tropical forest. We had passed many tributaries flowing into this lovely river, and finally we came to one that seemed especially inviting. It was dusk, and there was a good beach for our camp that night, so we stopped, made our camp fire, and decided that three of us would explore the tributary the next day using our small inflatable dinghy and the outboard.

Next morning we set off early up the tributary, its waters softly braiding their way down to the main river. It was gloriously sunny, and the trees on either side of us glowed a luscious green as they drank in the life-giving sunshine. There was a delicious intimacy here, as if the 'spirits' of this place were particularly benevolent, welcoming us into their secret world, far from the mad chaos of the modern world which was threatening the integrity of the great Amazon to the south. It was December 1976, six years since the completion of the Trans-Amazon highway.

After a few hours' travel we landed on a lovely golden beach, and explored the surrounding forest. It had a sacred hush about it—there was a powerful intelligence here which knew of our presence and which was glad that we had come. We spent a few quiet hours absorbing the subtle qualities of this place. Eventually, saturated with an unexpected grace, we climbed into the dinghy and set off downstream towards our camp, letting ourselves drift on the current without using the outboard motor. It was a gentle river without rapids—a few paddle strokes would be enough to steer us all the way back. We drifted without speaking, our senses alert to the sounds and sights of the vibrant forest.

Then we rounded a gentle bend in the river, and a whole new stretch of water opened itself to our enchanted gaze. A long way ahead of us there was a large fallen tree which had settled into the river, still connected to the shore by its giant trunk. The tree was a rich dark brown, and in the thick fallen branches that lapped the water we noticed a smudge of paler brown, difficult to make out in the distance. As we slowly drifted closer the pale brown smudge gradually resolved itself into a definite from—it was as if a smoky shape slowly coalesced itself into being out of nothingness.

It was Puma, one of the big cats of the forest, which had come down to drink its fill of the luscious river water. Puma stood drinking, totally unaware of us as we drifted silently closer. We blinked in amazement, for the forest seldom grants audience with the big tawny brown one who roams quietly and secretly through its great wild heart. It seemed as if time became thick, clear syrup pushing through us slowly, ever so slowly, and even though a sequence of events unfolded, they seemed to take place in some timeless realm which had for a few moments opened its doors to us.

Then Puma raised her magnificent, wild face and gazed upon us. A palpable sense of sinewy, relaxed intelligence rolled across the water towards us like tawny smoke, and we breathed it in, astounded by its profound quality. Drops of water fell from the short tufts of hair below Puma's mouth, sun-drenched diamonds which splashed slowly back into the river. Her gaze lasted a few seconds, but is there in us still to this day, for such blessings are eternal. Then, as we drifted closer, Puma slowly turned, and with exquisite grace, vanished into the forest.

It isn't just the tropical forests which emit cloud-seeding chemicals; the great temperate forests do it too, and so do the moss-covered peat bogs, and to a far lesser extent, the great northern boreal forests—the nearly continuous belt of coniferous trees that stretches across the far north of the American continent and Eurasia. In all, clouds seeded by life cool the planet by up to a staggering 10°C, about twice the temperature difference between a cold ice age and a warm interglacial period such as the one in which we are living now.

Since I learnt these amazing facts, I have seen clouds differently, through Gaian eyes. Once, they were for me the product of physics, now they are like fur or hair, not biologically alive in themselves, but the product of life nonetheless. They are the great dispersal buses of the microbial world that cool the planet, helping to keep it habitable overall, but also helping to trigger ice ages. Contemplating how most of Gaia's clouds are seeded by life gives us a taste of her animate presence, of her skill at handling the ever-brightening sun. Even clouds and the very wind itself are animate powers that living organisms have helped to set in motion.

Land and Ocean Working Together

We've seen how organisms on land and in the ocean help to regulate our planet's surface temperature, but could it be that these two great Gaian realms work together? To explore this question, Lovelock and his colleague Lee Kump made a model with land plants and oceanic algae coupled together. In the model, marine algae seed a dense sunshade of clouds, whilst on land great forests remove carbon dioxide from the air by enhancing the weathering of granite and basalt. These two great biotic communities cool the planet in tandem, but things change if a temperature increase is 'forced' through the gradual addition of carbon dioxide to the atmosphere. In this model, with 500 ppm (parts per million) of carbon dioxide in the air, the ocean temperature reaches 10^0C, and the marine algae vanish exponentially fast as a cap of warmer water develops on the sea surface. This starves the algae by preventing nutrient-rich currents, which sweep up from the sediments below, from reaching them. As the clouds above the ocean vanish and the dark blue sea is exposed to the warming rays of the sun, the planet warms rapidly to a new, hotter steady state, held there by the land plants working without the help of the algal denizens of the oceans.

In the real world, the cloud-seeding algae are retreating further and further polewards as human emissions of greenhouse gases warm the oceans. Perhaps the Lovelock and Kump model, for all its simplicity, captures something essential about the real-world climate system, for the flip to the new hot state happens when carbon dioxide in the air reaches 500 ppm—a conclusion also suggested by the latest generation of complex climate models.

It could also be that peat bogs on land and algae in the ocean work together to keep Gaia cool. The American scientist Lee Klinger has observed that DMS emissions from marine algae are highest in waters near coasts and islands. You would be right in thinking that this could be because nutrient-rich agricultural runoff, finding its way into the sea via rivers, gives the algae the nourishment denied them in summer time by the warm lid of water that develops on the ocean surface. This is exactly what happens in the heavily polluted waters off the coasts of northern Europe, but the effect also happens in places far away from agricultural contamination, such as in the Arabian sea, where algae thrive all year round despite strong seasonality in nutrient-rich upwellings. Klinger suggests that in the absence of upwellings, algae thrive best in coastal areas that receive river water from nearby peat bogs. As evidence, he cites the fact that massive plankton blooms in the St Lawrence estuary in northeastern Canada tend to happen when great surges of fresh water from rivers which drain peat bogs reach the sea. Clearly, there is something in the river water that the algae like.

The magic ingredient seems to be iron, which is scarce enough in the ocean to limit algal growth. Peat bogs consist of one major kind of plant—moss—most often only one kind: sphagnum. Mosses are happy in the wet and damp, and in order to keep things moist they release sulphur gases to the air which seed clouds and trigger rainfall. Like the rainforests, peat bogs maintain their own water cycle, and can kill large areas of forest by making things so rainy that the trees die of root drowning. Peat builds up when mosses on the surface of the bog die and are covered by new growth. The dead and dying moss releases humic and fulvic acids which inhibit decomposing bacteria so effectively that great masses of spongy black peat build up, which impede drainage, helping to kill trees and shrubs with the high soil moisture content. These peaty acids are adept at extracting heavy metals such as iron from the bedrock below the peat. The metals are *chelated* by these acids, and as those of us who take vitamin supplements know well enough, this means that the acid surrounds the metal in a protective chemical embrace which makes it much more easily absorbed across the cell membranes of living beings. Rivers draining peat bogs carry from two to ten times the average amount of iron, which reaches the algae in the ocean in easily digested, chelated packages.

The marine algae thrive on the peat bog's gift of iron, producing vast amounts of DMS just off the coast, not far from the peat bogs. The DMS (and other sulphur-bearing gases emitted by the algae, such as carbonyl sulphide, COS, and carbonyl disulphide, CS_2) seed clouds, some of which deposit sulphurous rain on the sulphur-hungry mosses sitting atop the peat bogs. In fact rain brings the mosses much more than sulphur, for they are perched so far above the nutrient-rich bedrock by the thick intervening layer of peat, that their root-like rhizoids have no hope of tapping into nutrients weathered out far below them, at the interface of rock and peat. The mosses thus have no option but to be almost totally dependent on rainfall for all of their essential minerals. If Klinger is right, the peat bogs and the oceanic algae feed each other the scarce nutrients each one needs, an unexpected case of cooperation between two great ecological communities. But this coupling doesn't only affect algae and moss; it also has major effects on the Earth's climate.

Peat bogs and marine algae cool the planet in similar ways. Both produce dense white clouds, and both absorb carbon dioxide directly from the atmosphere through photosynthesis. We've already seen how a rain of dead oceanic algal bodies settling on the sediments carry carbon sucked out of the air; this is the biological pump, but peat bogs also remove carbon dioxide from the air, fixing it as peat where it resides in dark, moist entombment for many centuries. Peat bogs also cool the planet by killing off dark, snow-shedding coniferous forest. Snow settling on the treeless bog creates a high albedo surface which cools the Earth about 80% more effectively than the snow-free moss of summer.

The bog–alga coupling is a positive feedback that could potentially plunge the Earth into a permanent snowball state if it moves in the direction of cooling. What prevents this from happening? Klinger proposes that as the world cools, advancing ice scrapes away the peat, reuniting it with oxygen in the air, giving off warming carbon dioxide gas. Even before the glaciers destroy them, the peat bogs could fix so much carbon that there is too little carbon dioxide left for photosynthesis, a negative feedback that would limit the growth of the mosses.

Biomes and Climate

Relationships between different biomes—Gaia's major ecological communities—also have a huge impact on climate. Up in the far northern latitudes, below the tundra regions where most of the peat bogs grow, lies the great boreal forest composed of dark evergreen pine trees which shed snow because of their triangular shapes. The trees' dark green foliage warmed the whole boreal region and the entirety of the northern hemisphere by a staggering 2.5⁰C during the period 1965–1995. This simple fact has a massive impact on the Earth's climate, for in the absence of trees the surface would be covered by snow that would cool the Earth. If tundra were to expand just a bit, North America and Eurasia would cool by around 3⁰C, and snow would lie on the ground for an additional 18 days.

There is a seesaw effect between the boreal forest and the mossy tundra which has played a major role in the earth's climate by amplifying the swings in and out of ice ages over the past 2 million years. As we shall see later, the trigger for these swings has been slight variations in Gaia's orbit around the sun. When our planet's elliptical orbit places us at our furthest distance from the sun, the tundra spreads its mossy legions southwards at the expense of the dark boreal forest and the snowy white winter tundra helps to tip the earth into an ice age. Thousands of years later, when our orbit brings us closer to the sun, the dark-leaved boreal forest expands northwards, warming the earth even in the winter months. The boreal forest also cools the earth by fixing carbon dioxide in wood, but the low albedo of its foliage warms the earth so much that this cooling effect is entirely cancelled out. All of this gives us a Gaian insight that undermines the idea that the tundra and boreal forest responded passively to climate as an external force. Now we know that the boreal forest, tundra and climate affect each other. The boundary between the boreal forest and the tundra is perfectly matched with the position of the boundary between cold northern air and warm air from the south known as the Arctic front. At first, it seemed as if the vegetation simply responded to where these two great air masses happened to meet, but we now realise that the position of the tundra-forest boundary controls the position of the front—a stunning demonstration of how far-reaching the climatic impacts of vegetation can be. Part of the reason for this is that the forest heats up much faster because it has a much lower albedo than the tundra.

But it isn't just the trees and mosses that interact with the climate of the high north; other members of the biotic community may also be involved, including predators and their prey. It is winter, and a wolf pack howls in the boreal forest in Isle Royale in Lake Superior, North America, signalling the start of another hunt. The wolves, like so many others in the far north, prey on moose, those giant ungulates whose males vie for females in the autumn by sparring with their widely spreading palmate antlers. Far to the south, over the Azores, warm air has built up during the hot summer, and now it has spilled northwards across the Atlantic Ocean rolling atop the cold air over Iceland, bringing high winds and deep snow to the entire western north Atlantic. Isle Royale has just experienced prolonged heavy snowfall, and our pack responds by combining with other wolves to make a big hunting group that will find it easier to bring down moose in the deep snow. The pack sets off to hunt at dawn. Today they will devour one large old male. Over the snowy winter months, our wolf pack, like many others on Isle Royale, and perhaps across the entire boreal forest, will feed many new pups with moose meat from kills facilitated by the deep snow. Fir tree saplings, browsed far less by moose, will grow better in the following year, increasing their chances of becoming mature trees that will warm the boreal region and the planet in many summers to come.

But not all winters at Isle Royale are as snowy as this one. Some years the pressure gradient that drives the massive northward flow of moist warm air from the south weakens, and little snow falls in the far north Atlantic. Then wolves hunt in smaller packs and find it more difficult to kill moose, which thus become more numerous. More fir saplings are browsed, reducing the numbers that eventually grow into large planet-warming trees.

There may well be more than a mere linear chain of cause and effect running from climate to wolves to moose to fir trees. The dark fir trees warm the boreal region and the planet, so the chain could well bite its own tail—there might well be a feedback. By influencing the abundance of moose and hence the cover of fir trees, it is not inconceivable that wolves could influence the temperature of the boreal region and hence the very pressure differences which lead to snowy or warm winters. The precise scientific details of this feedback, if indeed it exists at all, may defy elucidation. Perhaps a run of snowy winters favours hunting wolves and the growth of fir saplings so much that the trees shoot their way into the canopy in a decade or so, warming the region with their dark, snow-shedding branches. Once warmed, the temperature gradient down to the equatorial Atlantic may lessen, decreasing the chance of snow, thereby favouring the return of the moose as wolves hunt them less effectively in the now snow-free conditions. This begins to sound like a negative feedback with a long time constant, but no one knows. If it exists, the feedback could look something like this (Figure 29):

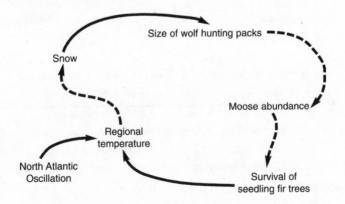

Figure 29: Hypothetical feedbacks between wolves and climate.

We don't have any answers, but by looking at the situation through Gaian eyes we can experience a delightful intuition of radical interconnectedness. There may well be a link between such apparently disconnected events such as the howls of a wolf pack, the very wind which ruffles their fur and the snow which gently covers the tracks of the stealthy quarry that they are setting out to hunt—all of this awakens a feeling of the astonishing wisdom that lies at the heart of our animate Earth.

Sensing Albedo

If you are in a warm, sunny climate or season, find a black cloth and a white one and place them next to each other on the ground, fully exposed to the sun's rays. After five to ten minutes, feel the difference in temperature between the two surfaces. If you can get hold of two T-shirts, one black and one white, wear each one in turn in the sun and feel how the albedo of the surface that covers your torso affects your body temperature and your level of comfort.

Now become aware of the albedos of objects and living beings around you. Pick a particular thing—a tree, a rock, a building—and spend some time connecting with it as a surface that reflects or absorbs the sun's rays. Based on your direct sensory experience of the black and white surfaces, become more and more aware of how your object affects the temperature of its surroundings entirely because of its albedo.

Now close your eyes and visualise Gaia as seen from space. Look at her dark areas, her boreal forests and the open ocean, and her light areas, the great banks of low clouds and the now shrinking areas of ice and snow. Connect with a feeling of how these dark and light areas make an important contribution to Gaia's temperature. See them changing as ice ages have come and gone, and as human activity increasingly perturbs the surface of our planet.

We have seen how climate and vegetation work together as a self-regulating whole. A further example of this in which local interactions predominate can be found in the Everglades in Florida, where the native pine woodlands have been cleared and natural marshes drained, with dramatic consequences. Simulations have shown that these drastic interventions have reduced rainfall by 11% because less water transpires from the trees and evaporates from both forest and marshland. With less evapotranspiration and less cloud cover, the treeless land surface has warmed by about 0.7°C—a catastrophically high increase.

The northern Sahara provides another example of the tight coupling between plants and climate. It is now a desert, but 6,000 years ago much of it was grassy bush country with scattered trees and shrubs, populated by a stunning assembly of large mammals, birds, insects and other wildlife. What could have brought about the shift from savannah to desert? Almost certainly around 6,000 years ago the earth's orbit brought us closer to the sun, triggering more rainfall in the region. The reasons for this are difficult to fully unravel, but it seems that changes to complex relationships amongst air, ocean, vegetation and sea ice in widely scattered parts of the globe were involved. Grasses and shrubs expanded their ranges in the slightly wetter climate, setting in train a positive feedback that encouraged even more rain to fall over the entire region. The ability of the soil to hold on to rainfall increased as plant roots permeated the once barren sand. As the plants grew, they sucked up water through their roots, releasing it into the air as water vapour that condensed into rain-bearing clouds with the help of cloud-seeding chemicals emitted by the plants themselves. The dark green surfaces of the plants encouraged the enhancement of evaporation. The feedbacks look something like this (Figure 30):

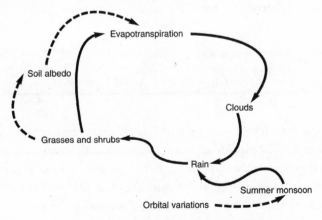

Figure 30: The drying of the northern Sahara.

About 5,500 years ago the rugged savannah country in the northern Sahara vanished unbelievably fast. Whatever happened set the feedbacks off in the opposite direction towards drying with exponential rapidity. Mathematical models suggest that a small change in the earth's orbit led to subtle changes in the distribution of sunlight over the earth's surface which were then amplified by long-distance teleconnections amongst sea ice, air, ocean and vegetation in various parts of the world, as well as by the more local feedbacks shown above. Modelling work combined with evidence from the field show how these slight changes conspired to push the system across a threshold beyond which the vegetation suddenly vanished (Figure 31).

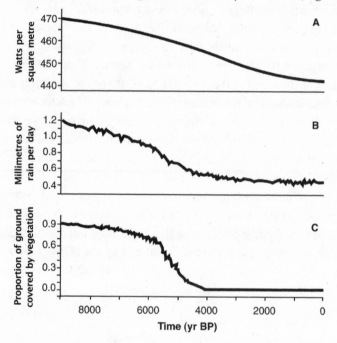

Figure 31: The rapid change from savannah to desert in North Africa 5,500 years ago. a: Change in the amount of sunlight received in North Africa, b: Simulated change in rainfall to the same region, c: Resulting change in vegetation cover. (*From Clausen et al, 1999: deMenocal et al, 2000*)

There's an important lesson in this story. It confirms our intuitions that small changes to Gaia's physiology can bring about dramatic unexpected changes, which is not surprising given that Gaia consists of innumerable interlinked linear and non-linear relationships. And there's a warning. By adding greenhouse gases into the atmosphere we could well trigger similarly sudden and dramatic changes—across the globe.

Chapter 7

From Microbes to Cell Giants

Gaia has evolved and changed throughout her long life, but these changes have not always been smooth and gradual—there have been critical transitions or tipping points that have reconfigured her atmosphere, oceans and land surfaces beyond all recognition, because of tightly coupled evolutionary developments amongst her living beings and her rocks, atmosphere and water cycle. The first such transition took place at her birth, when the first life forms, the ancient bacteria, spread all over the young planet some 3,500 million years ago at the beginning of the Archean eon. The bacteria ruled unaided until the beginning of the Proterozoic eon some 2,500 million years ago, when we are reasonably certain that the first nucleated cells appeared. Large multicellular beings composed of great colonies of nucleated cells did not arise until the beginning of the Phanaerozoic eon some 600 million years ago, during which the general trend, bar five mass extinctions, has been a gradual increase in biodiversity, culminating in today's climactic profusion of life. The key point to contemplate is that the appearance of new life forms in each eon changed things so much that an alien civilisation inspecting samples of air, water and rock taken at different times would have found it hard to believe that they could have come from the same planet.

During the Archean eon, long before there were multicellular beings such as trees, people and elephants, tiny bacteria ruled the world, as they still do today. Bacteria recycle gases and nutrients, and live in a vast variety of habitats: from deep, hot smoking volcanic vents at the bottom of the ocean, to bare rock surfaces exposed to wind, rain and intense sunshine. Ever since their appearance at the beginning of the Archean eon some 3,500 million years ago, bacteria have operated a tightly coupled metabolic network spanning the entire globe which has ensured that the earth's surface has remained within habitable bounds.

Bacteria Rule the World

To most people bacteria are nothing more than bringers of disease to be eradicated at all costs, but in fact only a small minority are pathological to humans. Most are engaged in the nitty-gritty, everyday work of keeping the planet alive by capturing energy, recycling and decomposing. But before we take a look at how bacteria do these marvellous things for Gaia, we need to learn something about them as individual beings. Let's shrink ourselves down so that we are only about 1/500th of a millimetre long, about the size of an *Escherichia coli* cell, a bacterium of about average size. Let's dive down into the soil, a place where bacteria teem in vast numbers; an average gram of soil contains millions of bacterial cells of many types.

There ahead of us in the gloom lies a single bacterium—a huge soil-straddling zeppelin. It feels squishy to the touch, like a water-filled balloon, for the cell's watery interior is bounded by a tough outer membrane that our puny sub-microscopic fingers can prod but not break. This is the microbe's cell membrane, studded with countless tiny pores through which even tinier molecules of food and waste constantly stream in and out. Brace yourself, for now we'll shrink down small enough so that we too can pass through a pore into the interior of the cell.

Here in the intense gloom we see a rich soup of molecular beings dashing about in a seemingly random frenzy. Snaking through them all like a twisted rope is the bacterium's genetic material, its DNA, which coils off into the distance like a giant earthworm plunging into the soil. We swim right up and tag the DNA rope with a red handkerchief and set off to see where it will lead us. As we move along the strand, our vista is unchanged—everything around us appears to be an undifferentiated mass of frenzied molecular movements. We travel for several hours, and there is no change in what we see, until there in the gloom we see a little red speck stuck to the DNA strand—it's our handkerchief! We've discovered that the strand is in fact a great *circle* of DNA.

We've learnt several interesting things from this brief journey to the bacterial world. The first is that the inside of our bacterium has very little structure—just about the only large organised thing we found was the long, circular rope of genetic material—the DNA. Furthermore, we found no organs of digestion, no gut, no stomach; so how does the bacterium eat and excrete? The answer lies in our other discovery—that the microbe's cell membrane is made porous thanks to the multitude of tiny holes that pepper its semi-fluid surface. The bacterium eats by absorbing tiny food molecules from its environment through these pores, and it excretes by sending wastes out through them. Life of any kind is utterly impossible without cell membranes, and all cell membranes display a stunning degree of agency and sentience, for all of them, no matter where they are found, are able to carefully select what goes in and out through their pores. We've encountered this fact before; cell membranes are *semi-permeable*.

Surprisingly, perhaps the most organised aspect of the bacterial cell is the chemical soup that we so glibly ignored in our first exploration. Let's shrink ourselves down again and return to the deep interior of our bacterium, but this time we'll stay in one place and look very carefully at the teeming chemical melée in front of us. After a while we recognise certain large molecular beings such as sugars, proteins and ATP, the universal energy-carrying molecule found in the cells of every single being, from microbes to elephants. After a few hours of careful observation and sampling we come to a stunning realisation: the overall composition of the cell's interior has remained relatively constant, despite the fact that the molecular beings are ceaselessly creating and consuming each other. This relative constancy is astonishing, for it suggests that the whole molecular network is able to regulate its own internal environment within limits favourable for its own existence. The cell does this by constantly re-making the components of the complex networks that operate inside its semi-permeable boundary. This continual minute-by-minute, second-by-second re-creation is a major hallmark of life, which the Chilean biologists Humberto Maturana and Francisco Varela have called *autopoiesis*—'self-making', or more literally, 'self-poetry'. In order for autopoietic entities to keep re-making themselves—in order to stay alive—it is indispensable that they have access to an external source of usable energy that can be chemically captured within the cell to be slowly released whenever and wherever it is required to build up molecules that have been worn down or have simply fallen apart during the rough and tumble of life itself.

But energy is needed not just for repair; it is also needed for eating and excretion. Living beings absorb nutrients from the outside world to stay alive: sometimes as tiny molecules brought in across a bacterial cell membrane, and sometimes as prey brought down by a pride of lions after a long hard chase in the intense heat of an African afternoon. After eating comes excretion. All life processes produce substances which would be toxic if not removed, and so all living beings must excrete, or better, donate, these substances to their surroundings to stay alive. The miracle of ecology is that the waste products of one kind of being are food for another, which means that living beings organise themselves into a vast, extracellular self-making network of which Gaia is the final expression.

In order to find new sources of food and avoid danger, some bacteria use an 'organ' known as the flagellum, a kind of whip which is used to move the cell around (Figure 32). A bacterial cell may have up to 400 individual flagella which reach out into the environment from every part of its cell membrane. When a bacterium such as *Salmonella* finds itself in a situation where a concentrated food source (say a rotting mouse in the soil) is leaching nutrients, receptors on its cell membrane detect the food molecules and instruct the flagella to rotate counter-clockwise. The flagella respond by tying themselves up into a single super-flagellum that beats in unison, causing the microbe to move in a straight line towards the food source. There is purposeful behaviour here—as the bacterium moves it keeps tabs on how many food molecules lock with its receptors, and movement towards the food continues if the number of 'hits' keeps increasing. Once the number of hits levels off the flagella rotate clockwise, causing them to separate so that each one beats to a different rhythm and at a different point along the cell membrane. The microbe then moves about randomly and with luck stays close to the source of food. In fact, both kinds of movement take place all the time, but straight-line movement dominates when the microbe is heading for food, with random movement happening most often when it is near food, or when an adverse situation is being avoided.

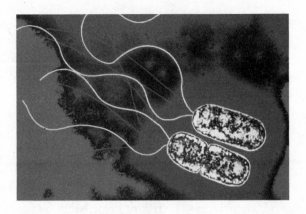

Figure 32: *Pseudomonas* sp., each with two composite flagella.
(*photo © Dr Linda Stannard, UCT / Science Photo Library*)

Bacteria also produce other kinds of structures. Some secrete a dense, felty mat of long, sticky fibres that grow from the cell membrane into the surroundings. These fibres, known as the glycocalyx, help the cell to stick to surfaces such as soil particles, bare rock and—in the case of *Streptococcus mutans*—to our teeth, where it causes tooth decay as long as the sugar sucrose is present to provide the raw material of the glycocalyx. Many different types of bacteria populate these sticky fibres in massive numbers, each type making its own unique metabolic contribution to the whole, and each able to communicate in highly sophisticated ways which have only recently come to light. We tend to think of bacteria as highly organised mindless matter, as nothing more than mere bags of molecular trickery— sophisticated no doubt, but far from sentient, and with nothing like the ability for sophisticated communication that is, we would like to think, the sole preserve of our own species.

But recent findings bring this conclusion into question. Emergence cannot take place without players that respond appropriately to the state of the whole, and bacteria are no exception. Most bacteria live in communities, often with different cell types carrying out specific metabolic functions, and in order for the whole to work well, the multifarious and multitudinous members of the group have to communicate with each other about the complexities of the surroundings which impinge on them, and about the state of the whole community. Response requires communication. In us, in our fellow mammals, in birds and in many insects, the major pathways of communication are sight, sound and smell. Amongst bacteria there is no sight, no hearing, but there is something akin to smell—the transmission of meaningful signals with which communication is possible, thanks to a versatile language whose alphabet is a complex set of chemical messages. One major channel of bacterial communication is known as *quorum sensing*. An amazing example lurks in the specialised light organs of certain squids.

We are snorkelling in a warm, shallow tropical sea on a bright moonlit night and occasionally catch sight of a strange glowing shape swimming furtively near the bottom. We've found a brave bobtail squid (*Euprymna scolopes*, Figure 33), which runs the risk of being eaten by predators should the moonlight project its shadow onto the sea floor. To avoid this danger, the squid engages in a cunning luminary deception: it emits light that matches the intensity of moonlight on the bed of the sea, thereby camouflaging itself from shadow-seeking predators.

Figure 33: A bobtail squid, *Euprymna scolopes*, whose internal light organs house bioluminescent bacteria.
(*photo © Margaret McFall-Ngai*)

But it isn't the squid that generates the light—bacteria known as *Vibrio fischeri* do this on its behalf. The bacteria live free in the sea, but, given the chance, love nothing better than to colonise the cosy vaults of a squid's light organs where, safe from their own predators, they consume a rich nutrient broth provided for them by their host. Whether in the sea or in a squid, each bacterium constantly produces small amounts of a signalling molecule known as AHL, which diffuses out of the cell and into the surrounding environment. All *Vibrio fischeri* cells are also capable of detecting AHL in their surroundings, using a receptor molecule known as LuxR. Out in the vast dilution of the sea, only small amounts of AHL pass from cell to cell, triggering no response in LuxR. But in the teeming bacterial populations inside the squid light organs, things are different. Here, each cell absorbs so much AHL from its neighbours that LuxR and AHL lock in a tight chemical embrace when the AHL exceeds a threshold concentration.

Together, the newly coupled molecules dance towards a specific region of the coiling bacterial DNA strand, bonding with it in yet another tight chemical marriage which sets into motion a beautifully choreographed chain of events ending in an astonishing feat: the emission of the eerie greenish light known as bioluminescence. This trick is not exclusive to squid; other animals have discovered that hosting dense light-emitting colonies of *Vibrio fischeri* can be of considerable benefit, such as the fish *Monocentris japonicus*, which uses them to attract mates. The light organs of squid and fish are the only places where the density of AHL becomes high enough to trigger light production. By its use, *Vibrio fischeri* recognises whether it is inside a light organ or in the open sea, and responds appropriately. It turns out that the use of AHL as a signalling molecule is not restricted to *Vibrio fischeri*—it is used by a wide range of bacteria, including some that cause diseases in humans (such as *Pseudomonas aeruginosa)* as well as nitrogen-fixing bacteria in the soil.

Another famous example of quorum sensing (this time by means of a molecule other than AHL) involves *Myxococcus xanthus*, a rod-shaped soil bacterium that forms flat gliding colonies in decaying vegetation. When things go well and there is enough food, *Myxococcus* cells divide and move along slimy trails in a tight pack, secreting prey-digesting enzymes in a stunning display of complex social predation described as "wolf-like" by one key researcher in the field. But when starvation threatens, things change dramatically. Under these circumstances the cells release a signalling molecule, which, above a threshold concentration, causes them all to converge into a great hump-shaped mound as they retrace their steps along their slime trails in a series of beautifully coordinated pulsing waves. Once in the mound, most of the cells commit suicide. The dying cells release nutrients which are used by a few survivors to make resistant spores able to sit things out until favourable conditions return. Here we witness 'mere' bacteria consulting with each other as a group. Each cell receives messages from all the others, reads its own internal state in relation to these messages, and then contributes its own response to the community pool of responses.

So far we have looked at how bacteria use a single signalling molecule for quorum sensing, but in fact most use several such molecules, and many are sensitive to signalling molecules emanating from species other than their own. *Vibrio fischeri* has a different signalling pathway which allows it to emit light when there are lots of other species in the light organ, so it must be sensitive to a whole host of signalling molecules. Furthermore, most bacteria use chemical signals to detect mutants that have become harmful to the colony. When this happens, the colony activates previously dormant genes, and becomes invisible to the mutants by, in effect, spontaneously switching to a different 'language'. One prominent investigator has said that these experiences with mutants help the colony to hone its "social skills", enabling it to improve its cooperation.

Communication amongst different species of bacteria allows them to form mixed species colonies which are able to accomplish tasks which a single species alone could never achieve. A startling example lurks in our own mouths, in the plaque that we so assiduously brush and floss away every day. Hundreds of bacterial species live in the plaque, their quorum sensing communication networks dwarfing the combined complexity of all our human communication systems. Next time you brush your teeth, give a thought to the sophisticated bacterial intelligence that you are so nonchalantly sweeping away into oblivion.

Bacterial chemical communication is of such startling complexity that it resembles the basic grammatical structures of human language, so much so that scientists are now talking about bacterial syntax and even about bacterial social intelligence. This sophisticated bacterial language allows for such tight coordination amongst different species in microbial colonies that they are best described as multicellular superorganisms. As in human language, the meaning of a given bacterial signal depends entirely on context, so that the same molecule will trigger a whole range of responses depending on what is going on both within and outside an individual cell. One key researcher in this field speaks of bacteria leading "rich social lives", of developing "collective memory" and "common knowledge", of having "group identity", of being able to "recognise the identity of other colonies", of "learning from experience", of "improving themselves" and of engaging in "group decision making", all of which add up to a social intelligence analogous to that of "primates, birds and insects".

What all of this amounts to is that we can no longer think of bacteria as nothing more than mere chemical mechanisms. Maturana and Varela hold that cognition is an indissoluble aspect of the self-making (autopoietic) quality of the living state, and since bacteria are undoubtedly autopoietic, their responsiveness to their inner environment and to the world around them are clear manifestations of a uniquely bacterial style of cognition. Bacteria are deeply sentient creatures that live in a rich, meaningful communal world, partially of their own making, to which they respond creatively and with exquisite sensitivity. The earliest spread of bacterial sentience around the globe some 3,500 million years ago led the nascent Gaia into an increasingly animate relationship with the brightening sun above, and with the carbon dioxide emitted into her atmosphere via volcanoes from the vast realm of semi-fluid rocks below. This great bacterial web has run the planet to this day, and is, in a way, rather like the unconscious processes that operate key aspects of our own metabolisms. There must be a seamless transition from this bacterial sentience to our own, for, as we shall see later, our very own cells are associations of once free-living bacteria that now engage in sophisticated intra-cellular communication. If our cells are fundamentally bacterial, then a continuous thread of sentience runs from us right back to our earliest bacterial ancestors.

Bacteria may appear to be simple when seen from the point of view of the huge multicellular creatures we are familiar with, but in fact they also display a communal metabolic versatility that dwarfs our own. Bacteria invented the major techniques for extracting and storing the energy needed for life, and for capturing key nutrients such as nitrogen and phosphorus at ambient temperatures. Very soon after life first appeared, they invented water-based photosynthesis, without which life as we know it would be impossible. They also invented fermentation, without which we would have no wine or cheese.

Bacteria have been able to succeed so brilliantly because of their immense capacity for networking, a skill they have been exercising through geological time, right up to the present day. A key networking skill that they operate in parallel with quorum sensing is the ability to swap bits of DNA, much as you or I sell or buy second-hand clothes at charity shops, only more so. Evolutionist Lynn Margulis and writer Dorion Sagan use a powerful analogy that beautifully illustrates this bacterial capacity for gene exchange. If, by some magic, your body could swap

DNA as bacteria do, copies of your DNA would leak out of you profusely as you swam around in your local swimming pool. If you were blue-eyed, the pool water would teem with multiple copies of blue-eye genes that had leaked into the water across your skin. Any brown-eyed swimmer plunging into the pool after you would easily absorb your genes and would emerge from their swim with startling blue eyes much like yours. Margulis and Sagan provide a further analogy. If we had the gene-swapping abilities of bacteria, then by merely smelling roses and inhaling the rose-smell gene we would smell like roses ourselves, or we could develop tusks just by spending a little time in close contact with elephants.

Oxygen-producing photosynthesis, one of the most astonishing of all bacterial metabolic accomplishments, may well have been invented by a single ancestral bacterium that spread the innovation by means of the 'open source' genetic exchange that we have just explored. We can illustrate this with a little story. Imagine that some 3,000 million years ago, during the Archean eon, there was a microbe named Suria who lived in the surface of the Earth's ancient ocean, swirling about in the great eddies which the wind made as it stirred up the waters. Life was already well established, and the air was full of methane, for this was long before the time when oxygen became abundant in the air. Suria, like many of her brethren, had been harvesting energy from the sun by splitting hydrogen sulphide gas from the depths into hydrogen and little globules of yellow sulphur, which she spat out into the sea. Then by some trick of the light, or by some remarkable coincidence of metabolism or genetics, a sudden but rather small change occurred deep in the tumultuous and beautifully orchestrated biochemical realm where the chemical beings which constituted her physical form played out their fantastic chemical intrigues, assassinations and love affairs. Suddenly, down in those depths, Suria's molecules configured themselves into a new way of dancing with sunlight and she was able to extract the life-giving hydrogen from water, H_2O, rather than from the pungent hydrogen sulphide gas all around her. Now she secreted oxygen rather than sulphur, and the world was forever changed. Suria, in creating 'oxygenic photosynthesis', had become the world's first cyanobacterium (Figure 34)—so called because of her novel cyan-coloured photosynthetic pigment. This was a fantastic achievement, and Suria, abiding by the microbial code of reciprocity, released small fragments of the genetic text bearing the secret of this new discovery into the sea around her. Soon, her immediate neighbours had taken up the message, and they too began to make oxygen and hydrogen from water. The recipe for oxygenic photosynthesis spread rapidly through the global microbial community, and very soon the superabundance of water made it possible for the photosynthetic cyanobacteria to abundantly populate many parts of the globe. Photosynthesis as we commonly know it had arrived, and has remained the major biological fixer of the sun's energy for the last 3,500 million years. The early alchemists had a great name for it; observing how green plants drink in sunshine, and sensing its power, they called it 'Green Lion Eating Sunlight'.

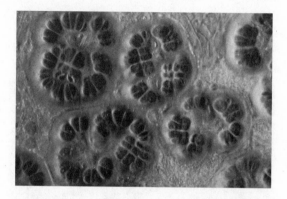

Figure 34: *Gomphosphaeria* sp., a modern cyanobacterium, from Laguna Figueroa, Baja California Norte, Mexico.
(*photo © Lynn Margulis*)

Some of the earliest photosynthetic bacteria lived in large colonies. Under the bright blue skies of Western Australia, in a place called Shark Bay, peculiar metre-high mounds litter the beach and foreshore like overgrown potatoes poking up through the sand (Figure 35). These are stromatolites—rock-forming communities of bacteria which were common all over the planet soon after life began about 3,500 million years ago. Mostly, each mound is a dome of rock, but on its top perches a cyan-coloured film of slime where the microbes live. Here, many species cohabit, each contributing to a veritable bacterial community as complex and communicative as any human city. Looking carefully, on top of a single dome we see silver bubbles appearing on its slimy surface. Some bubbles have already detached like miniature hot air balloons, and are dissolving in the crystal clear water. Freshly minted oxygen, made just as it has been for the last 3,500 million years, enters the atmosphere, energising the whole of Gaia. The bacterial community also creates the rocky dome on which it perches, either by secreting mucus that binds and lithifies particles of sediment, or by the direct precipitation of calcium carbonate. The bacteria constantly move towards the light at the top of their mound as it builds up, so that only the thin slime at the top is alive—perhaps in this way the bacteria keep themselves as close to the light as possible. Stromatolites are found today only in places where the sea water is so saline that bacterial predators can't survive, but in the long distant past of the Archean, 3,500 million years ago, these predators were absent and stromatolites were everywhere. Their significance for Gaia was their generous gift of oxygen, which gradually accumulated in the atmosphere, and their fixing of carbon-rich sugars locked into their microbial bodies.

Figure 35: Living stromatolites at Shark Bay, Western Australia.
(*photo © Reg Morrison*)

Yet the newly invented photosynthetic pathway could have quickly frozen the nascent Gaia to death by removing all the carbon dioxide from the atmosphere. This dreadful fate was prevented thanks to the efforts of the decomposing bacteria living in the sediments at the bottom of the ocean. These beings digested the corpses of photosynthesisers when they reached the sediments from the upper sunlit regions, releasing methane gas which bubbled its way upwards into the air, where its inordinate penchant for reflecting heat from the Earth's surface kept the planet warm.

At the ocean surface, wherever photosynthesis had made oxygen locally abundant, a new bacterial metabolic opportunity opened up: respiration. Respirers are extraordinarily cheeky, as they use oxygen to digest the very same photosynthesisers that produced the oxygen in the first place. In a sense, respiring creatures such as humans and the primeval bacterial respirers run photosynthesis in reverse, by using oxygen to break down the complex sugar molecules inside the bodies of their prey, releasing carbon dioxide and water in the process. Because photosynthesis and respiration balance more or less perfectly, oxygen cannot accumulate in the atmosphere if respiring organisms consume all the sugars fixed by photosynthesis. But what if some of the sugars escape the respirers and find their way into places such as thick, muddy sediments where there is no oxygen? Then the oxygen zooms about in the atmosphere seeking the lost sugars which, by eluding its attentions, ensure that the oxygen remains free in the atmosphere (Figure 36).

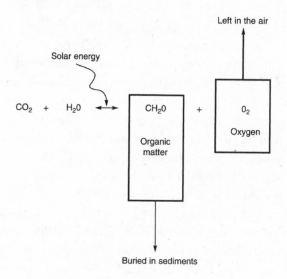

Figure 36: How oxygen builds up in the atmosphere due to photosynthesis and organic carbon burial.

In Gaia's early years, any oxygen that was not buried with organic carbon reacted with methane in the air, sulphur from volcanoes and iron in rocks. But eventually, around 2,500 million years ago, the very face of Gaia was gradually transformed beyond all recognition as these oxygen 'sinks' became saturated and free oxygen lingered in the atmosphere. Gone was the methane-rich air that had dominated the planet for 1,000 million years, gone the pink skies and brownish seas. Oxygen turned the sky blue and drove the methanogens deep into mud, ooze and slime where oxygen could not reach them. Gone too was the warm, balmy climate produced by the methane-rich atmosphere, for the planet plummeted down into an ice age as oxygen hunted down methane, converting it to water and carbon dioxide—which, molecule for molecule, is about 20 times less powerful a greenhouse gas than methane, its hydrogen-rich progenitor.

Bacterial Mergers

It was Lynn Margulis who championed the idea that the rise of oxygen may have fed back to trigger changes of massive significance amongst the denizens of the bacterial world. Like the sun and volcanoes, oxygen is both a giver of life and a dealer of death, for the gas is so highly reactive, so passionate in its quest to complete its outer orbit of electrons, that it happily attacks the complex molecules inside living bacterial cells. It could be that this new danger from oxygen prompted cells to develop a protective membrane to shield the genetic material in the nucleus, but in any case the new oxygen-rich environment quickened the global bacterial metabolism, allowing it to weather rocks on the land surface ever more effectively. As a result more nutrients were made available to life as a whole, strengthening Gaia's coupling between life and her realms of rock, air and water.

This newly enriched environment opened up possibilities for cooperative relationships amongst the bacteria that could only have been dreamt of by the inhabitants of the ancient methane-dominated world. Perhaps the first of these new associations took place between the highly mobile corkscrew spirochete bacteria and more sedentary bacteria such as *Thermoplasma* who lived out their lives in sulphur springs. Spirochetes are voracious predators, and at first may have remorselessly devoured beings such as *Thermoplasma,* but they soon discovered that a much more effective strategy was to attach themselves to the outside of their prey to suck up the nutrients leaking out of their cell membranes. Spirochetes are forever on the move, and so this arrangement meant that both partners were constantly propelled into new environments, some of which must have contained fresh supplies of food. The association worked as a stable configuration which enhanced the well-being of both partners. A similar association between free-living spirochetes and a host organism exists today in a most unlikely place—in the intestines of termites in the outback of northern Australia. These termites

(*Mastotermes darwiniensis*) eat wood—a major challenge, because the lignin in the wood cannot be digested by the unaided insect metabolism. Luckily for the termites, help is at hand in the form of wood-digesting microbes living inside their hindguts. One of the most spectacular of these beings, a large single-celled protist called *Mixotricha paradoxa* (Figure 37), provides the termites with a highly efficient wood digesting service. To do this, *Mixotricha* needs to move around in the termite hindgut, and this it does with the help of three types of spirochetes which cluster on its outer surface in vast undulating hordes, somehow orchestrating their individual movements so beautifully that the whole ensemble glides gracefully around in search of tiny splinters of wood.

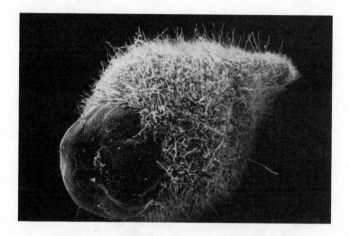

Figure 37: *Mixotricha paradoxa*, a wood-eating protist, swims only when the fuzz of its more than 200,000 motility structures undulate. These are hypothesized to have evolved from symbiotic treponema spirochetes.
(*photo © Dean Soulia, courtesy Lynn Margulis*)

Lynn Margulis has vigorously promoted the idea that when the Archean eon was drawing to a close, bacteria were trying out another kind of association, in which some of them lived inside other bacterial cells. This endosymbiosis, or 'symbiosis from the inside', may have begun as a predatory relationship which turned friendly once the predator discovered that cooperation suited it much better than naked aggression. The predators resembled the modern-day *Bdellovibrio*, a veritable tiger of the bacterial world which breaks its way into its victim, digesting it from inside using its high powered oxygen-breathing metabolism to produce vast numbers of freshly minted offspring that go off into the world as fierce hunters in their own right. But at some point some of these predators must have taken pause once they found themselves in the safe, nutrient-rich insides of their prey. Why go through all the bother of hunting down and devouring one kill after another, when it would be far more sensible to rein in the aggression and stay in this one nice host, feasting gently on its rich mix of food molecules whilst offering the host a useful service in exchange? And the service? Nothing less than the use of oxygen to extract energy from food molecules. This arrangement has been working successfully ever since it was first invented about 2,500 million years ago. Every single oxygen-breathing cell around today that isn't a bacterium is alive because of it, but the original bacterial predator has changed so much over time that its true identity was hidden from the prying eyes of scientists until relatively recently. The predator in its modern form is known to science as the mitochondrion (Figure 38)—the powerhouse of the nucleated cell.

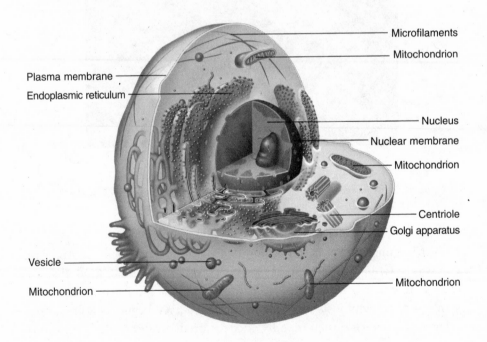

Figure 38: A nucleated cell showing its complex organisation,
including mitochondria, the descendants of ancient free-living bacteria.
(*photo © Science Photo Library*)

Mitochondria are about the same size as bacteria, but they don't look like them. Unlike the relatively undifferentiated innards of a normal bacterium, the insides of mitochondria are organised into a highly convoluted membrane rather like the high-hedged mazes one finds in the grounds of old English country houses. Excellent evidence has emerged from DNA sequencing that mitochondria were once free-living bacteria, for it turns out that mitochondrial DNA is much more closely related to that of bacteria than it is to the DNA in the nucleus of the host cell. The association between the mitochondria and the host nucleus has become so tightly coupled over time that the nucleus now produces some of the proteins found in mitochondria.

The last to arrive in the emerging symbiosis were the chloroplasts, once free-living photosynthetic bacteria that now live inside plant and algal cells as their photosynthetic powerhouses. The host cell must originally have been an oxygen-breathing symbiotic predator that probably derived a great deal of nourishment from eating the free-living, sugar-rich photosynthesisers. But here too there was a possibility for the transmutation of a predator-prey relationship into one of mutual benefit. For the predator, there were clear advantages in consuming some of the sugars made by the photosynthesisers, which in turn benefited from the protection and mobility of their host.

Journey to the Mitochondria

Find a quiet place to relax—your Gaia place perhaps—and spend some time breathing quietly. When you are ready, tune into the feeling of your own body. Dwell for a while in a sense of your body's warmth, of it resting on the chair or the ground, of your breathing, your heart-beat. Be aware of what you see, hear, taste and smell as you relax.

Now shrink yourself down smaller and smaller, until you become small enough to pass easily through one of the pores in one the multitude of cells that make up your liver. You are inside a liver cell, looking around at the stunning complexity and intricacy of the structures surrounding you.

Now choose one of the many mitochondria that float in front of you and move towards it. Touch it now, feeling its wondrous thrumming softness as it provides life-giving energy to the rest of the cell by breaking food molecules apart with oxygen brought to it from the air.

Sense the mitochondrion as an alien, as a being whose ancestors were once, thousands of millions of years ago, voracious free-living predators.

Savour the recognition that, long ago, the ancestors of this very mito-chondrion tamed their aggressive instincts and began to cooperate with the very cells they had once destroyed.

Feel the beauty and dignity of this 2-billion-year-old cooperative association. Sense how the supplies of energy liberated by the mito-chondrion allows your liver cell to carry out the many vital tasks keep you alive.

Now return to your normal size, and open your eyes. Look around at the more-than-human beings that surround you. Know how they, like you, are the descendants of ancient single-celled beings that learnt the subtle art of cooperation. The mitochondria teach us that independence is impossible—that we all depend on each other.

The Dance of Oxygen

By the end of the Archean era, the amount of oxygen in the atmosphere gradually increased as life became ever more adept at producing and burying carbon fixed from the air by photosynthesis. Amazingly, for the last 350 million years oxygen in the air seems to have hovered around the 21% that is close to optimal for large multicellular beings such as us. How has our animate Earth managed to regulate oxygen so effectively? It's a complex story, so we'll tease out only some of the details as they have been worked out by some of Lovelock's scientific descendants, namely Tim Lenton, Andrew Watson and Noam Bergman.

The first thing to note is that atmospheric oxygen cannot increase above 25% without triggering massive fires that would burn most of the land vegetation to ashes within a relatively short time. With this much oxygen in the atmosphere it wouldn't even matter if all the world's plant material was wet through and through—a global wildfire would burn it all to cinders anyway. If oxygen in the atmosphere declines to 13–15%, fires cannot start in even the driest vegetation, yet the continuous presence in the fossil record of partially burnt plant material, commonly known as charcoal, over the last 350 million years tells us that oxygen levels have never been low enough to prevent fires, nor high enough to totally burn up all the vegetation. The fossil charcoal hints at a compelling Gaian oxygen-regulation story involving surprising interactions between life, rocks, atmosphere and oceans.

The Flame of Life

Find a candle and place it unlit in front of you with a box of matches at the ready. Now take a match and light the candle, watching carefully as the flame leaps into life.

See the flame burning easily and constantly, and contemplate the fact that what you are seeing happens only because there is just the right amount of oxygen in the air. Had there been just 10% more oxygen, the flame created when you lit the match would have set you on fire, as well as the furniture in the room and then the whole house. From there the fire would have spread far and wide without stopping. Had you lived in South America, the fire would have spread over the whole continent and thence to Central America and eventually to the whole of North America. Anyone lighting just a single match on any island, or on any isolated land mass would have created a similar unstoppable fiery holocaust. On the other hand, with around 15% oxygen in the air, your brain would be unable to generate enough energy to sustain your consciousness, and you would be unaware of the candle, the flame or the fact that you are reading this text.

As you watch the candle flame, contemplate the miraculous con-
stancy of oxygen in the air that keeps the flame burning without
sparking a global wildfire, and that also keeps the flame of your con-
sciousness alive. Become aware of your breathing—take in great lung-
fuls of oxygen, and contemplate how its presence at just the right
concentration is the great gift of the astonishing relationships between
rocks, the living beings such as lichens, trees and mosses that weather
them, oceanic algae, bacteria in the ocean sediments, and that rare
and life-giving chemical being, phosphorus.

Phosphorus, one Gaia's most important chemical beings, is at the centre of the story. Two key needs of living beings give it this centrality. It is absolutely indispensable for making the energy-storing molecule ATP in all living beings, which is also needed for growth and as a component of DNA. But there is another crucial fact: its scarcity. The ultimate source of phosphorus is the weathering of rocks, and its ultimate destination, once weathered, is the ocean sediments. There is no gaseous phosphorus-bearing chemical being that can transport phosphorus from sea to land, no chemical equivalent of the gas dimethyl sulphide that so beautifully completes the sulphur cycle by raining sulphur onto forests, bog and fen; and so living beings are totally dependent on the cycling of rocks for fresh supplies of the precious element—the gold of the biological world. Contact between any acid and rocks is all that it takes to weather out their precious lodes of phosphorus. Carbonic acid from the marriage of rainwater and carbon dioxide does the job well enough.

Two things happen when tectonic movements have uplifted huge amounts of phosphorus-bearing rocks, rich in organic carbon. Firstly, oxygen in the air decreases, consumed by the oxygen-hungry organic carbon. Secondly, chemical weathering releases phosphate ions from their rocky prisons to be swept away by rivers to the sea, where they stimulate the growth of phytoplankton that release oxygen and fix organic carbon. (An important digression: each phosphate ion consists of a phosphorus atom covalently bonded to four oxygen atoms. As pure phosphorus is not found in nature, in what follows I will, for simplicity's sake, use 'phosphorus' and 'phosphate' interchangeably.) Oxygen increases, as long as the organic carbon is buried away in the ocean sediments. Herein we can see a simple negative feedback that regulates oxygen. But there is more to the story, because there is another source of phosphorus. The sediments on the continental shelves are rich in phosphorus that has come from the weathering of rocks by a number of different routes. It seems likely that in the sediments live phosphorus-hungry bacteria that need oxygen in order to hoard away the precious element like so much looted treasure. With little oxygen in the sea water, the sediment bacteria have little aptitude for capturing phosphorus, and so it remains free in the sediments. Phosphorus can also enter or leave the sediments thanks to a powerful but purely chemical relationship with iron. In oxygen-rich waters, the two chemical beings feel such a strong attraction for each other that phosphorus is 'sorbed' by iron, but in oxygen-poor waters the relationship breaks down and the phosphorus is released back to the sediments. Some of the free phosphorus that has not been taken up by bacteria or by iron has a chance of becoming incorporated into rocks as the sediments slowly settle and harden, but some escapes this rocky fate and is swept up to the surface by ocean currents that feed new generations of photosynthesising algae. And so oxygen increases. Now negative feedback begins to counteract this increase: with more oxygen in the water, the iron-phosphorus marriage and the sediment bacteria fix more phosphorus in the gloomy depths, and so the surface algae starve. With less photosynthesis and hence organic carbon burial, oxygen in the air decreases. But there is a fatal flaw in the story, for the feedback collapses if oxygen reaches even moderate levels in the water just above the sediments, because the crucially important switching between phosphorus capture and release vanishes.

The most effective oxygen regulation journey involves the enhanced weathering of phosphorus-bearing rock by land plants. We've already seen how plants amplify the chemical weathering of granite and basalt by fracturing and dissolving the rocks with their roots and potent chemical exudations. A key discovery made by Tim Lenton was that the same life-enhanced weathering liberates significant amounts of phosphorus from calcium silicate rocks and from sedimentary rocks replete with organic carbon. This is his new feedback. If oxygen in the air increases significantly, fires rage everywhere as the vegetation on the vast land surfaces of the planet burns, leaving huge swathes of country with severely reduced plant life. Fewer land plants weather less phosphorus out of the rocks, and the whole biosphere suffers from phosphorus starvation. Photosynthesis is severely reduced in the phosphorus-poor world, and so the burial of organic carbon diminishes on land and in the oceans. As a result, oxygen levels in the atmosphere plummet as oxygen-hungry chemical beings such as carbon, iron and sulphur gobble up their highly reactive quarry. But now, with less oxygen in the air, the danger of fire recedes and the land plants grow back vigorously, releasing phosphorus once again as their roots probe, crack and dissolve the rocks in their compulsive search for nutrients. The newly liberated phosphorus stimulates photosynthesis on land and in the oceans, and the world's green beings increase the oxygen content of the air as more of their carbon-rich bodies are buried in the murky sediments. The atmosphere once again teems with oxygen, and the great self-regulating dance comes full circle as fires once again rage over the continents.

We've now explored three major negative feedbacks that may have regulated oxygen over the last 300 million years or so. Each is a Gaian journey, and all are linked, as shown in Figure 39. Spend some time tracing the three journeys in this diagram until you are familiar with how each one works to regulate the amount of oxygen in the air. If you find it helpful, refer back to Figure 8 to remind yourself of what the two kinds of arrow mean (a solid arrow indicates a direct relationship between two components, and a dashed arrow an inverse relationship). Once you have a good cognitive grasp of Figure 39 as a 'systems diagram', you are ready to experience its deeper significance by shifting your attention from the cognitive to the experiential mode of awareness. The first thing to do is to transform the systems diagram into a map of oxygen regulation that we will use for the 'journeys' below.

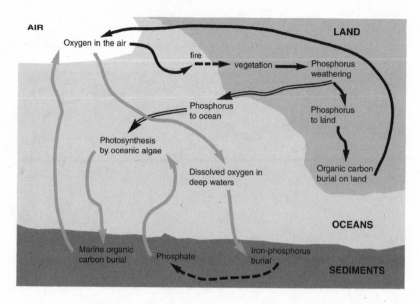

Figure 39: Three oxygen-regulating journeys.

Three Oxygen-regulating Journeys

For each journey, the words within the brackets tell you where to begin, or which is your next stop (ignore them once your familiarity has increased). We will travel around the first journey to begin with, so now spend a little time giving the following map some special attention.

Go to your Gaia place, and place the map of the first journey nearby, so that you can open your eyes to glance at it whenever you need to. Take a few deep breaths, and allow yourself the time and space to take in the sounds, smells and sights of your Gaia place.

The First Journey

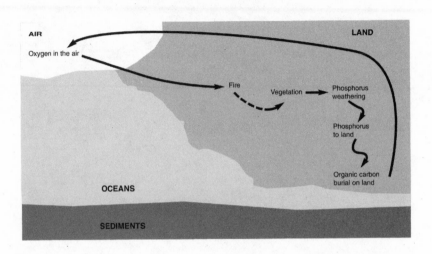

(Begin at 'oxygen in the air'): Breathe gently, 'tasting' the oxygen that you take into your body with every in-breath. Savour the life giving oxygen molecules as you draw them deep into your lungs. Feel the oxygen coursing through your body, giving you energy and igniting your consciousness.

You are an oxygen molecule swirling through the atmosphere, feeling your corrosive yet passionate double-edged quality as both a giver of life and a bringer of death. You rush wildly on a gust of wind that almost blows you onto a cliff face rich in oxygen-hungry organic carbon molecules that had once been part of a tree fern in a Carboniferous forest 300 million years ago. Many oxygen molecules vanish here, but you are blown away from the cliff at the last minute by a passing breeze. There are so many oxygen molecules in the air that great wildfires rage all over the planet, set by lightning strikes.

(To 'fire'): You take the form of a wild Jarrah tree in the great bush country of south-western Australia. Fires rage nearby in the oxygen-rich atmosphere, and soon the trees around you are alight. Smoke billows around you, and fleeing animals rush past you on all sides. Feel the heat of the fire as it comes towards you, searing every living thing in its path.

(To 'vegetation'): The great fires have burnt away vast swathes of the forest, but you are one of the few surviving trees. Most of the bush around you is black with the charred remains of what was once a rich and diverse ecological community.

('To 'phosphorus weathering'): Your roots dig deep down into the granite, cracking open the rock, breathing out carbon dioxide that combines with water molecules stored away from rains long past. The carbonic acid that results from this union dissolves the granite, releasing a host of long-incarcerated chemical beings, including precious phosphorus. But the fires have killed so many trees that now the forest as a whole liberates only small amounts of this rare element from the rocks.

(To 'phosphorus to land'): You become a rare phosphate ion freshly weathered from the granite by the great Jarrah tree. Rain washes you into the sparse soil, but there are no other phosphate ions near you. You are quite alone.

(To 'organic carbon burial on land'): You become the great wild being of the entire Jarrah forest, extending over 3.9 million hectares of south-western Australia, a great network of trees and plants connected by thin threads of underground fungi. But in the phosphorus-poor conditions very few of your plants grow well, and most are small and stunted. There is less photosynthesis, and so your soils bury far less dead plant material.

(To 'oxygen in the air'): Completing the journey, you once again take the form of an oxygen molecule in the air. Organic carbon burial on land has decreased so much that there are hardly any new oxygen brothers in the air to replenish those that have combined with so many oxygen-hungry chemical beings. The journey you have just completed has diminished oxygen in the air, averting a catastrophic global wildfire.

Now quickly travel around the journey once more to experience how it springs into action to avert another danger—the loss of too much oxygen.

You become Gaia. Now, with less oxygen in the air there are fewer fires, and you feel the land vegetation growing lush and green once more over your land surfaces. Plant roots weather huge amounts of phosphorus from the rocks, and your plant cover grows even more. As the burial of dead plants continues apace in the phosphorus-rich world, you taste the tangy legions of oxygen molecules increasing in your atmosphere.

The Second Journey

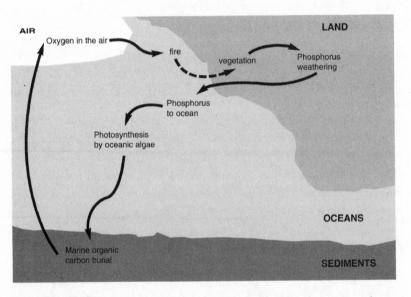

In this journey, the events leading from 'oxygen in the air' to 'fire' to 'vegetation' to 'phosphorous weathering' are the same as those you have just experienced on the first journey. Taste the oxygen in your in-breath as we begin at 'phosphorus weathering'.

('Phosphorus weathering' to 'phosphorus to ocean'): You are an ion of phosphate weathered out of the rocks by the nutrient-hungry land plants. The great fires on land triggered by abundant oxygen have killed off much of the rock-weathering vegetation. As a result, very few phosphate ion brothers share your journey into the tumbling river that washes you down over boulders and cataracts to the sea.

(To 'photosynthesis by oceanic algae'): You are now a single-celled diatom gently floating on the surface of the sunlit ocean off the coast of Africa. Your whole being trembles with an insatiable hunger for phosphorus that the sea cannot fulfil. Now you become all the tiny drifting light-gatherers in the world's oceans—each one a single photosynthesising cell. You feel the collective hunger of the entire ocean's free-swimming photosynthesisers for the phosphorus denied them by the massive forest fires on the nearby continents. The hunger is so great that it stifles photosynthesis and growth.

(To 'marine organic carbon burial'): You become the murky sediments at the bottom of the ocean, exquisitely sensitive to the soft touch on your muddy surface of dead beings sinking down from the upper sunlit regions of the sea. Far fewer carbon-rich dead bodies have come to you in recent times because of the great fires that have raged way above in the realms of forest, air and light.

(To 'oxygen in the air'): The decreased organic carbon burial at sea means that there is now less oxygen in the air, preventing the proliferation of wildfires that could have wiped out much of the life on land.

Now travel around the second journey once again to feel how it works to replenish oxygen. Becoming Gaia, you feel how with less oxygen in your air land plants once again grow thickly on your continents, weathering huge amounts of phosphorous from your rocks. Your sinuous, richly braided rivers offer much of the life-giving phosphorus to the ocean algae, helping them to grow abundantly. Multitudes of their dead bodies sink into your sediments, far from the ardent attentions of oxygen, which now remains free and abundant in the luminous air.

The Third Journey

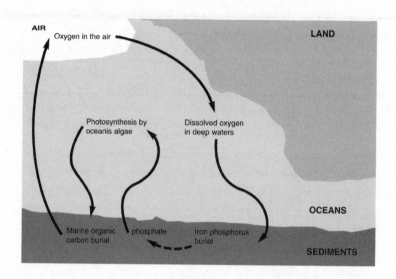

Open yourself to the sounds, sights and smells of your Gaia place. Feeling deeply at peace, take a few deep breaths and once again taste the presence of oxygen in the air.

(Begin at 'oxygen in the air'): You are an oxygen atom swirling about in the atmosphere at a time when oxygen has increased to danger- ously high levels. You feel a sense of overcrowding as you bump into many other oxygen brothers, all intent on passionately combining with almost any chemical being they meet in their wanderings.

(To 'dissolved oxygen in deep waters'): You are blown far out to sea by a fierce gale that is churning the North Atlantic into a great mass of foam and swirling water. Suddenly a wave catches you, and you feel its cool, watery embrace as you dissolve into Gaia's ocean realm. Here everything is much slower, much more settled than in the air. You feel the great calm that comes from being held so closely by this huge body of moving water.

A great downwelling tongue of water carries you into the ocean deeps along with many other oxygen brothers gleaned and dissolved from *the air. Feel the immense cold tranquility of the deep ocean as you gradually approach the sediments.*

(To 'iron-phosphorus burial'): You sense the dark, oozing mud engulfing you as the cold water finally delivers you to the sediments. Total darkness surrounds you. A nearby iron molecule awakens to your presence, and with great excitement links you and itself to a nearby phosphate ion, preventing its escape into the eager currents of sea water that constantly brush the top layer of sediments.

(To' phosphate'): You become one of the few phosphate ions that have managed to escape the ardent attentions of the iron and oxygen beings in the sediments. You settle at the surface of the sediments in the dark gloomy depths of the ocean, a precious but scarce essential nutrient.

(To 'photosynthesis by oceanic algae'): Powerful ocean currents carry you upwards towards the sunlit regions of the sea. Here, myriads of tiny light-loving beings seek to devour you. Soon, a passing diatom engulfs you and you become all of the marine algae floating in the sunlit regions of Gaia's oceans. You feel the intense hunger of all your tiny photosynthetic beings as growth and reproduction are greatly diminished in the phosphorus-poor waters.

(To 'marine organic carbon burial'): You transform into the dark sediments at the bottom of the ocean. The rain of dead algal bodies reaching you from the upper sunlit regions of the sea has greatly decreased because of the lack of phosphorus. You miss the gentle touch of their arrival on your murky surface.

(To 'oxygen in the air'): You become a rare oxygen molecule released into the air by a photosynthesiser that soaked up sun and a few scarce phosphate ions at the ocean surface during its brief life before it was buried in the sediments. Sense how the air is less crowded now with oxygen brothers. Many have been consumed in chemical marriages with organic carbon exposed on the mountain slopes, and with sulphur and other gases released by volcanoes. The great self-regulatory dance has reduced oxygen in the air.

Now quickly travel once more around the journey to experience how it prevents a calamitous oxygen decline. Becoming Gaia, you sense the diminished oxygen in your atmosphere. Now there is less dissolved oxygen in your deep ocean waters, which allows more phosphorus to escape the ardent attentions of iron and enter the sea from the sediments. Algae now grow well in your sunlit ocean surface, nourished by the abundant phosphorus. As more algal dead bodies rain down into your sediments, more oxygen is left behind to circulate freely in your swirling air, powering the biosphere with its gift of life. The great oxygen-regulating dance has come full circle.

Rest calmly now, gently breathing the oxygen-rich atmosphere that makes our life possible. Tasting oxygen once again, feel your indissoluble connection to our animate Earth.

Lenton and his students are continuing their quest for what could be called a 'Gaian web of life'—a model linking the ocean- and land-based feedbacks we have just explored with the stories of other key chemical personalities such as sulphur and nitrogen. Nitrogen is a key nutrient that is made available to life only when it is removed from the atmosphere by bacteria in the soil or in the open sea. But the nitrogen-fixing bacteria cannot do their work without phosphorus. So if oxygen in the air decreases, land plants spread and weather more phosphorus into the ocean, which stimulates the growth of nitrogen-fixing bacteria. Some of the newly fixed nitrogen finds its way into the photosynthetic marine algae, so oxygen increases as more dead algal bodies sink down to murky tombs in the ocean sediments. This is another neat negative feedback that is tightly linked in with phosphorus. Another group of bacteria, the denitrifiers, strip nitrogen out of dead bodies, returning it to the atmosphere to complete the nitrogen cycle. Amazingly, the quantitative relationships between phosphorus, oxygen, carbon dioxide, nitrogen and the living beings in the model result in an emergent self-regulatory dance that has kept all of these chemical beings well within the limits that life can tolerate over hundreds of millions of years.

Another key insight from this work is that Gaia's ability to self-regulate may well have improved as life has discovered new modes of being and self-expression. Around 440 to 420 million years ago Gaia experienced a fundamental transition when land plants appeared and took hold of her land surfaces. Before this there was little oxygen in the atmosphere, and the cycles of carbon dioxide and oxygen had very little to do with one another, but the cycles of these key chemical beings became tightly linked after the spread of land plants. This happened because of the heavy dependency of land plants on both oxygen and carbon dioxide: too much oxygen destroys land vegetation with fire, whilst carbon dioxide is a key nutrient that stimulates plant growth. It is amazing that land plants play such a major role in setting the levels of the very gases that so deeply influence their own growth. Plants increase oxygen through photosynthesis and organic carbon burial, and by weathering rocks rich in phosphorus; and they decrease carbon dioxide by amplifying the weathering of granite and basalt. Plants also remove carbon dioxide from the atmosphere through photosynthesis. In so doing, a new self-referential network has spontaneously emerged that acts to regulate oxygen much more tightly with land plants than without.

Chapter 8

Desperate Earth

If Gaia speaks, what does she tell us? I'm writing in a small, quiet room in a meditation centre on the edge of Dartmoor, and have just come back from a walk amongst the rocks, streams and open moorland of this wonderful open country in the south-west of England. As I gazed out over the dun coloured hills before me on a brilliantly sunny afternoon on the last day of January, a gentle, cold wind blew over the barren land. Where once, thousands of years ago, there was a vast, windswept forest, now there is only open moorland peppered here and there with a few lonely trees. What happened? The rocks and the thinly scattered trees have the answer: they tell of the first Neolithic agriculturalists who came here over 4,000 years ago with their stone ploughs and their fierce, sharp stone axes. The great granite tors recall how these men cut down the trees to plough the light soils for their crops. They remember the fires and the felling, the new fields and the foreboding as the axes cut deep into the flesh of millennial trees. The stones speak of this even now, and so do the few lonely trees that have been lucky enough to escape the endless pressure of fire and over-grazing. Despite 4,000 years of abuse, the land still gives us beauty and grandeur, for there is still great presence here—the wild consciousness of a place diminished, but not yet fully desecrated.

Dartmoor speaks a warning, for what happened here so long ago is now happening all over the planet. There is now no part of Gaia left untouched by the human hand, not even the furthest reaches of Antarctica or the deepest depths of the ocean, for the contaminated atmosphere connects everything and reaches almost everywhere on the Earth's surface. Everywhere 'development' chews up wild places, spitting them out as the 'stuff' we increasingly see as indispensable for our lives. How will Gaia respond to this onslaught? According to James Lovelock's latest understanding (outlined in his 2006 book *The Revenge of Gaia*), the answer will almost certainly be abrupt, catastrophic climate change, which will increase global temperatures to levels not experienced on the planet for at least 55 million years. The destruction of New Orleans by hurricane Katrina in September 2005, as well as many other recent serious climatic events, are a sign that we have unleashed Gaia's wrath, and that in her desperation she seems poised to respond to our onslaught with an even greater one of her own which will kill vast numbers of people and lay low our so-called civilisation.

Gaia in her Natural State

So what was Gaia's state of being before we began to disturb her—what was she up to? How can we even attempt to answer such questions? Some of the answers are to be found in Gaia's diverse modes of speech: as bubbles of ancient air locked up in polar ice, as different versions of her chemical beings laid down in rocks, and as the signs left by ice, wind and water on the very rocks themselves. These are Gaia's memories, some faint, some vivid, but for the most part amazingly coherent over space and time. These memory traces tell us that for most of her life Gaia has gone through million-year periods of relative warmth, and equally long periods of cooler temperatures. But things have changed in the more recent past. We know for certain that for the last two million years Gaia has been moving in and out of ice ages with extraordinary regularity. Every 100,000 years over the last 700,000 years ice has spread down from the Arctic regions into the northern temperate latitudes, covering them with glaciers kilometres thick, and the world has cooled. We have already seen in Figure 5 how the oscillations are stunningly regular, as regular as the heartbeats of Gaia's own living animals, but with a very slow pulse, the sort of pulse that could only belong to a great whale of a planet coursing silently through space, a pulsing back and forth from ice to warmth in a cycle about fifty times as long as the entire span of Western civilization. Periods of ice have dominated most of the time, but every 100,000 years, for a relatively brief spell, Gaia has experienced a time of warmth before plunging back down into an icy state (Figure 5). Carbon dioxide and methane, two of Gaia's key greenhouse gases, have pulsed in step with these changes in temperature, strongly hinting at their central involvement. But there is a further message in these traces that we have already encountered: Gaia's temperature has reached the same maximum at each warming, the same minimum at each cooling. These tight bounds, beyond which she hasn't strayed for the last two million years, speak of the presence of a tightly coupled self-regulating dance between life, rocks, atmosphere and ocean.

The recent ice ages have to do with the fact that Gaia, although apparently alone in the vastness of space, is deeply sensitive to the animate presences of the other planets in our solar system, and of course to the heat stress she is experiencing from the sun, which is now hotter and brighter than ever before. Gaia engages in conversations with the community of planets; she is privy to a larger realm of interactions, mediated by that mysterious attraction to which we so glibly ascribe the word 'gravity'. There, out in the vastness of space lie the great gas giants, Jupiter and Saturn, and closer than us to the sun lie Venus and Mercury. And nearby, at a mere stone's throw by galactic reckoning, is our own moon. Each is an animate presence, a character in its own right, and each expresses itself in part by drawing our planet closer to its own ambit, much as an Aboriginal lover in the Great Sandy Desert beckons to his beloved using the ancient technique of *yilpinji,* or love magic.

These conversations and attractions cause Gaia's orbit around the sun to vary from perfectly circular in three important ways. First there is eccentricity of her trajectory though space, which expands and contracts from egg-shaped to almost circular with a periodicity of 100,000 years. Then there is the tilt of her spin axis, which returns to a maximum of 24.5 degrees from a minimum of 22.5 degrees every 41,000 years. Finally, there is precession, referring to the top-like wobble of her spin axis that takes about 25,700 years to complete. Currently, the North Pole points towards *Polaris*, the North Star, but 5,000 years ago it pointed to a different North Star, *Alpha Draconis*, observed in wonder by the ancient Egyptians from their pyramids.

These orbital variations determine the distribution of solar energy reaching our planet. Precession and tilt by themselves cannot alter the absolute amount of solar energy reaching Gaia's surface, they only alter its distribution, but the changes in eccentricity do make a difference in this regard because the absolute distance of the Earth from the sun is altered. Even so, the amount of extra energy received is tiny—only 0.2% less at maximum eccentricity relative to the almost circular minimal eccentricity.

The fact that the periodicity of Gaia's orbit matches that of the swings from glacial to interglacial conditions is taken as evidence that eccentricity is the main driver of the cycle. But how can such a small change in the amount of solar energy reaching the planet make such a difference? The answer seems to be that a suite of global-level positive feedbacks are acting as amplifiers.

According to James Lovelock, the fact that such a tiny change in solar energy can have such drastic consequences is a sure sign that Gaia's temperature regulation is in crisis—that she is experiencing some kind of pathology. As we have seen, during the last 3,500 million years Gaia has maintained habitable temperatures in the face of an ever-brightening sun by gradually removing carbon dioxide from her atmosphere, thanks to the ever more efficient life-assisted weathering of granite and basalt. By two million years ago, so much carbon dioxide had gone that removing any more under a modern and very bright sun has been of little help in Gaia's efforts to keep herself cool. Inventors know only too well that cybernetic systems wobble from one state to another when they are about to fail, just like a wildly oscillating top before it runs out of energy. For Lovelock, Gaia's recent wobbles in and out of ice ages may be a clear sign that she is struggling to keep cool under a bright sun—that she is overstretched to the point of instability, with the glacial periods being her preferred state, and the interglacials her fevers during which she hovers dangerously close to catastrophic climatic breakdown. Without the small addition of solar energy given her by the eccentricity of her orbit Gaia would probably have oscillated somewhat randomly between ice ages and interglacials in her search for a comfortable coolness. If so, Gaia has driven her own shifts from cold to warm, but the rhythmical expansion and contraction of her long trajectory around the sun has imposed regularity on what might otherwise have been a far less predictable plunging in and out of ice ages.

So how does a tiny increase in solar radiation trigger the end of an ice age? What are the amplifiers? A hundred thousand years of ice have passed, and Gaia's orbit once again contrives to bring her marginally closer to the sun. The dark oceans warm as they soak up the tiny bit of extra solar energy. Molecules of carbon dioxide, methane and water, picking up speed in the warmer water, leave the oceans to journey in the vast expanses of the atmosphere as greenhouse gases, further amplifying the warming. The ice, once safe in the grip of an ice age, is not immune from the warming. At last, after 100,000 years of torpor, huge numbers of its water molecules are liberated into the fluidity of the liquid state. As the melting continues, more and more of the dark land and ocean are exposed to the sun, and Gaia warms further. In the northern hemisphere, the boreal forests advance, swallowing up the peat bogs, warming the region and the entire world as their dark, snow-shedding leaves absorb the sun's warmth and as the dying peat bogs release their immense stores of carbon to the air.

The additional warming affects the oceans, which release even more carbon dioxide into the air. Plants grow well in the new high carbon dioxide atmosphere. They send their roots deep in search of nutrients, cracking open rocks with sheer brute force and with the subtle but relentless dissolving powers of their acidic chemical exudations. One can almost hear the gentle grinding noise of the increased weathering as plants all over the planet pummel and pulverize the rock, releasing nutrients on a scale unknown during a time of ice. Myriads of phosphorus, iron, silicon, calcium atoms are captured by plant roots to be sucked up into the growing green biosphere which, in its heedless growth, draws out more and more carbon dioxide from the atmosphere. But some of the newly liberated chemical beings manage to avoid the terrestrial green world altogether, and, tumbling headlong into rivers and thence to the sea, are put to work in the photosynthetic dance of the minuscule phytoplankton that populate the ocean. In time, over thousands of years, Gaia's photosynthetic beings draw out so much carbon dioxide that the great efflux from the oceans is counteracted and she experiences a brief interglacial with a maximum carbon dioxide concentration of 280 parts per million.

But this warm state is a rare, transient thing. It cannot last, for the power of photosynthesis on land and in the ocean, in their immensity, has taken so much carbon dioxide from the air that the emissions from the warm oceans are totally absorbed, and more. With less carbon dioxide in the atmosphere, Gaia once again begins her descent into an icy world as the positive feedbacks begin to move the world towards cooling, aided now by the gradual and subtle lengthening of her orbit. In a cooling world, the oceans absorb more carbon dioxide, driving temperatures down even further. The boreal forests retreat, giving ground to the cooling, carbon-hungry peat bogs, covered in winter by a reflecting blanket of snow. Favoured in the colder world, ice from the far northern hemisphere begins to spread south again, further chilling the planet with its whiteness. Although seriously reduced in the high latitudes, the great biotic communities on land expand in the south-east Asian tropics, where the lowered sea level exposes new terrain almost as large as Africa. Here rainforests grow, reducing carbon dioxide even more, whilst carefully garnering precious hordes of phosphorus atoms thanks to their intricate rock-weathering skills. The ground waters and rivers swell with this newly found mineral wealth, and swirl it down to the oceans where thankful phytoplankton bloom in a last exuberant frenzy of photosynthesis and cloud-seeding. Now the cooling world dries out, as less water evaporates from the oceans. Great pressure differences between the topics and the high latitudes stir up strong winds that carry iron-rich dust from the drying land far out to sea, bringing a further boon of nutrients to the shimmering, temperature-reducing phytoplankton. As the northern ice expands, the global ocean circulation reconfigures itself into its contracted mode, cooling the planet even more. Gaia has drifted into a comfortable new age of ice, with around 180 parts per million of carbon dioxide in her atmosphere until, 100,000 years later, the ellipse in her orbit curves her once more into the right alignment for warming.

The fact that Gaia has switched from ice ages to interglacials on a regular basis might lead one to think that her climate has remained stable in either of these two states, but nothing could be further from the truth. Cores of ice from central Greenland are much better at revealing climate changes on short time-scales than the ice cores from Antarctica because of the way in which the Greenland ice was deposited. The ice from the north holds unequivocal memories that climate during both the last glacial and the current interglacial was highly unstable, with particularly rapid shifts from relatively warm to cold even when the world was in the grip of the ice age. The periods of warmth were significant, with the world warming within decades to temperatures close to those of today, followed by a slower cooling back down to glacial conditions. The evidence suggests that these temperature fluctuations were triggered by very small changes in solar luminosity—another warning that seemingly insignificant changes can be amplified beyond all recognition into huge effects by complex dynamical systems such as Gaia.

Scientists know only too well that irregular behaviour is a hallmark of complex systems. One particular event underscores this instability: the end of the Younger Dryas cooling some 11,600 years ago, when global temperatures soared by 15^{0}C in no less than a decade. Why this happened is not totally clear, but emissions of methane either from wetlands or from the huge store of undersea methane hydrates were almost certainly involved, together with the reorganisation of the circulation of the global ocean. The messages from the ice are strikingly clear: stable climate is a myth. The Gaia into which our species emerged is a wild, complex dynamic being, subject to sudden shifts between multiple semi-stable states. At this time in her long life, small disturbances can ramify through her vast body, growing larger and larger through positive feedback, as our explorations of the effects of tiny increases in solar luminosity have so clearly shown. There are tipping points beyond which climate can suddenly transmute from benign to deadly, and there is no good reason for us to bask in the complacent idea that our emissions of greenhouse gases will warm the planet gradually—that we will have time to adapt. It is far more likely that we will trigger abrupt, catastrophic climate changes that will push Gaia into a new hot state unsuitable for many of her life forms, including ourselves.

Gaia and the Western World

So how are we changing Gaia's climate, and what are the likely consequences? The scientific community has addressed this vital question in part through the work of the IPCC—the Intergovernmental Panel on Climate Change, a coalition of thousands of scientists from around the world. Their work involves data analysis combined with super-computer simulations of Gaia's past, present and future climate. In their last report, known as the TAR (the Third Assessment Report) published in 2001, the IPCC stated that human emissions of greenhouse gases are already having a discernible effect on climate, and that doubling the carbon dioxide content of the atmosphere as compared to pre-industrial levels would produce a temperature rise between 1.4^{0}C and 5.8^{0}C, a situation that will probably be reached at some point during this century. Commentators have noted that this amount of warming would constitute "the biggest temperature increase in the history of civilisation".

Of course, we have already drastically changed our climate, as was indisputably confirmed in February 2005, in the week when the Kyoto Protocol was at last ratified, when American and British scientists announced a dramatic warming of the world's oceans over the previous 40 years that could only have been caused by our pollution of the atmosphere with greenhouse gases. Other data show that global average temperatures have increased by about 0.6°C during the 20th century—the warmest century for a millennium. Such a small increase may not sound very important, but it masks much greater regional temperature changes. Furthermore, the warming has happened unbelievably fast compared to previous pre-industrial changes, even though the biotic communities of both land and ocean have absorbed about 60% of our emissions in roughly equal measure. The mean global temperature graph from the TAR covering the period from 1000 to 1999 shows a clear downward trend until about 1900, when the planet's temperature began to climb rapidly to today's high level. This suggests that before our interference Gaia was headed for the next ice age, possibly in about 15,000 years. Greenhouse gas emissions have risen in step with the temperature increase. We now have about 380 parts per million carbon dioxide in our atmosphere, about 30% above the ceiling of 280 parts per million to which Gaia has returned during each of the previous four interglacial periods. There is now a general consensus that a doubling of atmospheric carbon dioxide from pre-industrial levels to 550 parts per million could happen by 2050, and that this would increase global temperatures by 2°–4°C.

Recently, researchers harnessed the spare processing capacity of computers owned by members of the general public in order to carry out many more climate simulations than were possible for the IPCC. The most likely temperature increase for a doubling of carbon dioxide that came out of these additional simulations was around 3.4°C, somewhat similar to the IPCC predictions, but the shock came when researchers found a much wider range of outcomes than did the IPCC, with a small number of outlier simulations (the most extreme results) showing temperature increases ranging from 2° to 11°C for a doubling of carbon dioxide. The implications of this new work are that the high-end predictions of severe warming, although not very likely, must be taken very seriously.

The key effect that the researchers varied in this new work was the behaviour of clouds, which have been called the 'Achilles heel' of climate models because they are notoriously difficult to represent mathematically.

When carbon dioxide warms the air, more water evaporates from the oceans, so in a warmer world we would expect to find more water vapour in the atmosphere. Water vapour is a powerful greenhouse gas in its own right, so increasing it could lead to even more evaporation and hence to even higher temperatures—a positive feedback on warming. But things aren't that straightforward, because water takes on many guises as it journeys around our living planet. It can manifest as liquid water, or as a solid, as in ice and snow, or as ten major kinds of cloud, each of which can have an overall warming or cooling effect both regionally and globally, depending on the altitude at which the clouds form, how long they last, and how white they are. For example, those wispy cirrus or mare's tails that form high up on clear days may well be overall warmers, whilst the dense low marine stratus clouds, which give Britain its well deserved reputation for virtually unending gloom, are coolers. What clouds will do in a warmer world is still a mystery, but as the warm-end outliers of new simulations show, there is a chance that clouds will disperse sooner under higher temperatures, or that the few clouds that do manage to appear will quickly drop their rain and vanish. Both scenarios would increase the warming feedback as more sunlight reaches Gaia's surface and as the greenhouse effect of the additional water vapour delays the escape of infrared radiation to space. There are some alarming indications from the real world that these sorts of things are already happening in the tropics, where skies have become less cloudy since the late 1980s. No one is certain that this effect is due to climate warming, but many climate scientists suspect that there is a connection, and many are concerned that the IPCC climate models don't cope reliably with clouds. There seems to be an emerging consensus that the models with the most unrealistic cloud feedbacks are the ones that produce low-end results, so these may well be excluded from the next IPCC report due out in 2007. Another serious limitation of the climate models is that they may never be sufficiently able to represent the climatic influences of living beings, such as the seeding of clouds by marine algae and vegetation on the land.

Despite these uncertainties, the TAR predicted that the human community would experience severe disruption as a result of climate change, and that there would also be very negative consequences for the biodiversity of our planet. The increasing concern prompted the British government to convene a major international conference at Exeter in February 2005 entitled 'Avoiding Dangerous Climate Change'. The findings of this conference are alarming. In the words of the Steering Committee report, the conference agreed that: "Compared with the TAR there is greater clarity and reduced uncertainty about the impacts of climate change across a wide range of systems, sectors and societies. In many cases the risks are more serious than previously thought." Conference participants deemed that "increasing damage" was likely, with temperature increases of 1^0–3^0C, and that there was a risk of "serious, large scale system disruption" once average global temperature increase exceeded 2^0C. Many experts think that this level of temperature increase will take place at some point during this century when atmospheric carbon dioxide levels reach around 400 parts per million. When this happens, Gaia could move though a series of irreversible tipping points, such as the melting of the Greenland ice cap, the re-configuration of the global ocean circulation, the disappearance of the Amazon forest, the emission of methane from permafrost and undersea methane hydrates, and the release of carbon dioxide from soils. Add to this the insidious effects of climate change on biotic communities around the world, and the enormity of what we are doing to our planet becomes shockingly apparent. What are the implications of crossing each of these tipping points?

In the far north, the ice glints in the summer sun, and the Arctic feels warmer than it has for thousands of years. The 2-kilometre thick Greenland ice cap is already melting, at a rate of about 10 metres per year, ten times faster than was previously contemplated—40% of the sea ice around the North Pole has melted over the last 33 years. The melting is happening faster than previously imagined partly because of the infamous ice-albedo feedback. A small initial warming in the long Arctic summer melts some ice, exposing the dark surface of either land or sea, which warms the region even further, thereby promoting even more warming, and so on. The loss of Arctic sea ice by September 2005 was so severe that scientists now think that the far north has reached an irreversible tipping point that will lead to no sea ice in the far north within a century. As the sea ice vanishes, the entire region warms, increasing the melting of the Greenland ice cap. But there is another positive feedback particular to the Greenland ice cap that also contributes to its demise—the ice cap loses height as it melts and so progressively encounters warmer air above it. If all of the ice on the Greenland ice cap melted, sea levels around the world would rise by seven metres within 1,000 years, with catastrophic impacts on civilisation, which is centred mostly in vulnerable coastal cities. And it won't take much of a temperature increase to make this happen, as the tipping point seems to be 2.7⁰C—well within the TAR predictions. Moreover, once the melting has started in earnest, nothing can stop it, not even a drastic fall in carbon dioxide emissions. To add insult to injury, the melting of the Greenland ice cap would send enough fresh water into the sea to flip the oceanic thermohaline circulation into its 'off' mode very quickly indeed, perhaps triggering changes to the climate of western Europe and possibly beyond.

But it is not just Greenland that is melting—so is the vast majority of Gaia's realm of ice and snow. As the great melt proceeds, Gaia experiences positive feedbacks on warming. In the last 30 years, the extent of snow cover in the far northern hemisphere during spring and summer has decreased by 30%, and even though there was a slight increase in winter snow cover in North America, the albedo decrease of the exposed bare ground in the warmer months has made a net contribution to the overall warming of the region and the planet. The disappearance of sea ice in the Arctic has encouraged explorers to look for a navigable route through the long sought-after Northwest Passage between Greenland and Canada. Polar bears are threatened with extinction, for they will no longer have vast expanses of sea ice over which to roam in their hunt for seal. Caribou, moose and polar bear—the future for these animals looks bleak as their domains become increasingly blocked by expanses of open water or thin ice.

The realm of ice and snow is also under siege in Antarctica, where dramatic changes are taking place as the continent warms at a rate of about 0.5⁰C per decade. The 31st January 2002 is a date to remember, for on that day large sections of the Larsen B ice shelf began to break away from the main mass of the icy continent. Huge amounts of ice were involved—32,500 square kilometres, an area bigger than Rhode Island weighing in at around 720,000 million tonnes, broke away in a 30-day period, littering the surrounding ocean with icebergs of all sizes. This is an unusual event. Recent evidence shows that Larsen B has been stable for the last 10,000 years, so its demise can be linked with some confidence to our warming of the climate. Ice shelves sit on the ocean and snuggle right up against Antarctica like whale calves alongside their mothers, so the collapse of an ice shelf doesn't by itself lead to an increase in sea level. Ice shelves are made of ice sliding down into the sea from glaciers on the continent, and seem to act like 'corks in a bottle', preventing the glaciers behind them from plunging catastrophically into the ocean. Remove the corks, and the 'unstopped' glaciers surge into the sea, unleashing dramatic rises in global sea level as their huge bulk adds to the overall volume of the world's ocean. There is a real danger that the Ross ice shelf could soon begin to collapse. This ice shelf is the size of France, with 200-metre-high cliffs at its seaward end, and a maximum depth of one kilometre. Its break-up would cause so much ice to reach the sea from the West Antarctic ice sheet that global sea levels would increase by 5–7 metres. Indeed, the melting of the entire Antarctic ice sheet would raise global sea levels by as much as a phenomenal 50 metres. These and other changes to the distribution of ice shelves are playing havoc with Antarctic wildlife. Penguins are suffering greatly as they are cut off from their traditional feeding and breeding grounds. Further assaults on Gaia's icy realms are taking place in the world's glaciers, which are extremely sensitive to climate change. Most glaciers decreased in length during the last century, but a few, notably those in Norway, have expanded. The observed retreat of global glaciers is consistent with a global warming of between 0.46^0 and 0.86^0C over the last 100 years.

Gaia's icy realm also extends into the great regions of permafrost up in the high northern latitudes, where the subsoil has remained frozen for several thousand years to depths of up to 1,600 metres. Vast areas of our planet's land surface, about 25% of the total, are covered by permafrost, including half of Russia and Canada and 82% of Alaska. Every summer the uppermost layers of permafrost melt in the sunshine, but as climate change takes hold the melting goes deeper and the permafrost slowly retreats northwards. Permafrost holds huge amounts of organic carbon (about one-seventh of the world's total), mostly in the dead bodies of mosses and other plants, and its melting gives microbes access to huge amounts of organic material—food to them. As they binge on this vast bonanza, the microbes release carbon dioxide and methane to the atmosphere, contributing another twist to the multifarious and accumulating positive feedbacks that are warming our planet.

Much of the permafrost forms in gravelly or silty soils held together by ice, and when this melts the ground becomes slushy and deforms. In parts of Alaska, this thaw-induced subsidence has caused deformations of up to 2.5 metres, and with it the collapse of buildings and pipelines. The loss of the permafrost has severe consequences for the traditional peoples of the high Arctic, who find it much more difficult to hunt in the soggy ground, and their prey, notably the reindeer, also face difficulties because their migration

routes are altered by the melting ground. The thawing in the far north is also
changing biotic communities as fens and bogs invade birch forest.

But another, possibly greater danger lurks in the permafrost and in
some of the continental shelves of the world's oceans. In such places,
where temperatures and pressures are just right, a remarkable association
takes place between two of Gaia's key chemical beings—water and
methane—giving rise to the increasingly infamous 'methane hydrates'. In
these curious structures, water abandons its usual penchant for making
hexagonal ice molecules and assembles itself instead into curious cubic
cages of ice with up to eight 'guest molecules' of methane in cordial repose
at the centre of each water cage, made up by no less than 46 water molecules. The methane comes, of course, from the activities of decomposing
bacteria which live, as they have since Gaia's earliest days, in sediments
poor in oxygen but rich in the dead bodies of creatures from the upper
airy, sunlit reaches of the planet. Apart from an abundant supply of
methane and water, low temperatures and high pressures are absolutely
indispensable for the appearance and survival of these flimsy, delicate
molecular cages. Where temperatures are low, pressure is less important to
their survival, and methane hydrates form in cold, waterlogged terrestrial
environments such as in the permafrost. But at higher temperatures, great
pressures are needed to keep them intact, and the hydrates can only be
found under a great weight of water such as along continental margins
where there are enough dead bodies to feed the hungry methanogenic bacteria. Here, methane hydrates can reach a thickness of up to 500 metres.

Huge amounts of methane have been stored away in hydrates over millions of years. The total store in the hydrates and in free methane gas
trapped beneath them is in excess of 10 million (10,000 billion) tonnes, by
far the biggest reservoir of organic carbon on the planet and about 13
times as much carbon as is held in today's atmosphere. Huge amounts of
carbon are stored in the permafrost hydrates alone—almost as much as is
held in the sum total of all the world's terrestrial biotic communities. If
burnt, methane hydrates would yield more than twice the energy held in
all the world's combined reserves of oil, coal and gas.

Herein lies the danger, for methane hydrates easily fall apart when they
experience slight changes in temperature or pressure. When this happens,
methane is released to the atmosphere, warming the planet. It is almost
certain that the breakdown of methane hydrates helped to warm the Earth
at the end of the last ice age, and they may even have contributed to the

Permian mass extinction, when most marine creatures died, perhaps because of suffocation as huge amounts of free methane reacted with oxygen. We would be wise to anticipate similarly drastic methane releases in our times, due to our frenzied emissions of planet-warming gases. We don't know quite where the tipping points for catastrophic methane emissions are, but there are serious suggestions that a warming of 3^0C would lead to the release of 85% of the methane after a few thousand years— effectively an irreversible change, as far as humans are concerned. The melting is happening as you read these words. Reports just in from the permafrost of western Siberia, one of the fastest-warming regions of the planet, reveal that millions of hectares of frozen bog are melting, suggesting that a vast release of methane will most likely shortly follow. A quarter of the total methane stored on land, some 70 million tonnes, could be released from the region if its permafrost melts completely.

There is another terror associated with the dissolution of methane hydrates: tsunamis. Stable hydrates are as solid as rock, but as soon as they fall apart, previously solid substrate turns to liquid mud that creates a sub-sea slump that can whiplash the water into coherent killer waves. The third giant Storegga slide on the Norwegian continental shelf is an example—7,000 years ago the tsunami it created dumped sediments four metres above the high tide line in parts of Scotland.

A relatively new and incontrovertible discovery discussed at the Exeter conference was the acidification of the oceans that comes about when the carbon dioxide we are pumping into the atmosphere dissolves in sea water to give carbonic acid, which in turn releases hydrogen ions, those minuscule, highly reactive chemical beings whose increasing presence in a liquid defines it as more acidic. The newly added hydrogen ions seek out and remove carbonate ions in the sea water, a loss that eventually forces calcium carbonate to restore carbonate to the ocean by dissolving itself—an act of ultimate chemical self-sacrifice that has dire consequences for the numerous living beings that so deftly use chalk for making shells, coccoliths and skeletons. These include the corals, the crabs and sea urchins, the clams and sea shells, and the tiny but climatically vital coccolithophores so critical for taking carbon out of the atmosphere on a variety of timescales and for seeding planet-cooling clouds. As the chalky structures of these diverse creatures dissolve in a global ocean enriched with hydrogen ions put there by our lust for burning of fossil fuels, Gaia warms even more in yet another destructive positive feedback.

Meanwhile, down in the Amazon basin, things aren't looking too promising either. The endless sea of green forest, so vast and seemingly immune
from any serious human intervention, is also approaching a critical tipping
point. Global temperature increases may well see to it that the forest rapidly vanishes soon after 2040, when its ability to recycle water suddenly
collapses after the region has warmed above 4⁰C. The great expanse of forest, whose canopy, seen from the cockpit of a light aircraft, looks like an
ocean of billowing green clouds stretching off towards the horizon, will be
gone, replaced by the yellow hues of savannah vegetation with its scattered
trees and drought-tolerant grasses. The dying Amazon will release its vast
store of carbon to the atmosphere, as bacteria and fungi feast on the dead
bodies of its giant trees and on its once luscious vegetation, and as fires sear
their way through the landscape. This carbon, in the form of carbon dioxide and methane, will join the burgeoning legions of carbon atoms in Gaia's
atmosphere, further warming our world. The demise of the Amazon will
change global climate—the wheat belt in the mid-United States could suf-
fer less rainfall, with drastic consequences for world grain supplies. Vast
areas of carbon-rich tropical vegetation around the world are also burning,
particularly those in south-east Asia, where farmers set fires to clear peat
bogs on Borneo and Sumatra for their crops, releasing large amounts of
planet-warming gases to the atmosphere.

Soils hold particularly large amounts of carbon—globally about 300
times as much as we release every year by burning fossil fuels. It is autumn
in England, and the great horse chestnut tree at Schumacher College is
shedding its great, spread-eagled golden leaves in preparation for the coming winter. On dry days, the gardeners come to take these leaves away for
compost making. Eventually, the compost is put out on the flowerbeds in
the famous Dartington Hall gardens. Much of the carbon in the compost
will stay in the soil—a tiny contribution to the great global soil carbon
reservoir. But with rising temperatures, decomposing bacteria will swarm
more vigorously and more abundantly through the soil in their ceaseless
search for food and nutrients. Wherever they find it, they will consume the
soil carbon, releasing carbon dioxide and methane back to the atmosphere. Normally, photosynthesis sends more carbon from the atmosphere
into the soil than is released by decomposition, especially in an atmosphere enriched with carbon dioxide, but this tidy arrangement is predicted
to tip and turn tail when carbon dioxide levels in the atmosphere reach

about 500 parts per million sometime within the next 10–50 years, as carbon-absorbing photosynthesis is overwhelmed by carbon-releasing decomposition. Alarmingly, researchers have recently shown that we may have already tipped over this threshold. To the surprise of experts in this field, soils in England and Wales during the period 1978 to 2003 have vented carbon to the air at an alarming rate, not because of land use changes but almost certainly because of climate change. This is a significant loss of carbon that almost exactly cancels out Britain's technologically achieved reductions in carbon emissions from 1990 to 2003. The same thing is almost certainly happening in many other parts of the world, and as soils become a net source of carbon to the atmosphere, the great planet-warming positive feedbacks receive another turn of the screw, as higher temperatures release more soil carbon, sending temperatures spiralling even higher.

Sensing Climate Change

Sit or lie on the ground in your Gaia place, taking in the sounds, sights and smells around you. Relax deeply, and breathe easily.

Imagine that your taste buds have developed an extraordinary sensitivity for carbon dioxide molecules in the air. Spend some time breathing gently, developing your skill at sensing these vitally important chemical beings as they spin and eddy in and out of your body.

Now use your newly acquired skill to taste the vast numbers of new carbon dioxide molecules emitted into the air by our industrial culture.

Feel the presence of these carbon dioxide molecules in your lungs, just as they are now more present in Gaia's air, oceans and living beings. Taste the huge numbers of carbon dioxide molecules sent into the air over the last 150 years by our burning of coal, oil and gas, and by our destruction of forests and peat bogs.

Species on the Move

Perhaps the clearest evidence that climate change is a reality comes from what is happening to the world's biotic communities, which are being affected around the world in a variety of alarming ways. There are changes in phenology, which refers to the study of the timing of key events in the world of living beings, such the precise dates and seasons of flowering or egg laying. Then there are changes in the overall distribution of species, including whether their overall ranges are expanding or contracting. Next are changes to the composition of biotic communities, and in the interactions between these species. Lastly there is concern about changes to the overall functioning of whole ecosystems, and the impacts on the 'ecosystem services' they give us.

In the temperate regions of the world, spring-time events such as flowering, budding, singing, spawning, migrant arrival and insect appearances are not what they used to be—they have all been happening progressively earlier since the 1960s. In Britain alone, 16% of flowering plants flowered significantly earlier during the 1990s compared to previous decades, and similar trends are being noted all over the world. There are also changes to some key autumnal events, such as amongst some migratory birds, which have abandoned winter migration altogether, or have delayed their departures.

Each species has its own very specific range of tolerance for temperature and moisture, and species move in an attempt to live within their comfort zones as the climate changes. The general trend in a recent study of 1700 species is a poleward movement of 6.1 kilometres per decade, and a 6.1 metres movement up the sides of mountains. Virtually the whole biosphere is being uprooted in unprecedented ways. Examples are legion, including the northward march of the boreal forest at the expense of open tundra vegetation; the northwards expansion of red foxes in Arctic Canada and the simultaneous shrinking in the range of the Arctic fox; the upward movements of alpine plants in the European Alps by 1–4 metres per decade; the increasing abundance of warm-water species amongst the zooplankton, fish and intertidal invertebrates in the North Atlantic and along the coasts of California; and the extension of lowland Costa Rican birds into higher areas from lower mountain slopes, because of changes in the frequency of dry season mist. Butterflies, too, are being disturbed (they respond very quickly to climate change); in Britain and North America, 39 butterfly species have moved northwards by up to 200 kilometres in 27 years.

But many of these exiled species face another danger—habitat fragmentation. As economic growth and development proceed apace, more and more of Gaia's wild places are paved over, built on, or ploughed up for intensive agriculture, preventing species that need to move to higher latitudes and altitudes from reaching suitable safe havens. As these species face extinction, biotic communities lose complexity and diversity and unravel further.

In a rapidly changing world each species is out for itself, and as a result many biotic communities begin to fall apart. In Britain, smooth newts (*Triturus vulgaris*) all over the country feel the urge to dive into their breeding ponds earlier and earlier as temperatures climb, just as soon as there is the slightest sign of spring. This is unfortunate for the common frog (*Rana temporaria*), which does not respond to the warming climate in a similar way. By the time the frogs arrive to breed, the ponds are full of plump, well-developed newt tadpoles who love nothing better than to dine on the defenceless frog embryos and newly hatched frog tadpoles. So the common frog, already decreasing because of pollution and habitat destruction, is inched closer to oblivion by a secondary effect of climate change. The impacts of its loss on the wider ecology are as yet unknown—perhaps British gardeners will notice more slugs eating their lettuces.

A similar unravelling of biotic communities is going on all around the planet. It is early spring in the Hoge Velue, the Netherlands' largest nature reserve, and great tits are hunting for their staple food at this time of year—the caterpillars of the winter moth, that in turn feed on newly emerged leaves of the oak tree. For millennia the great tits (*Paris major*) all over temperate northern Europe have timed their breeding to coincide with the extraordinary abundance of winter moths (*Operophtera brumata*) in early spring, but for the last 23 years the great tits in the woodlands of Hoge Velue haven't been able to find enough winter moth caterpillars with which to feed their chicks. The culprit is climate change. Winter moth females lay their eggs high up in the canopies of oak trees, and the caterpillars must emerge when oak leaves are good to eat just as they are unfurling from their buds. If the caterpillars emerge too soon, there will be no leaves and they will die of starvation; and if they emerge too late, the caterpillars face the prospect of eating oak leaves full of indigestible tannins. Winter moth caterpillars use ambient temperature to time their emergence, and over countless generations this method has paid off—but no longer. With increasing spring temperatures, the caterpillars are hatching up to three weeks before oak bud-burst. They survive only a few days, but the great tits at Hog Velue have not learnt to bring forward the timing of their own breeding to take advantage of this brief glut, so they too are facing severe food shortages at a critical time of year. These sorts of 'phenological decouplings' are happening all over the planet as a consequence of climate change. We know little about what the impacts on biotic communities will be, and even less about how these impacts will affect the feedbacks between the biological realm and climate.

Hurricanes and Global Dimming

So far we have not yet considered any significant negative feedbacks that might oppose the warming trend. There is one important effect that we haven't mentioned yet—global dimming. It turns out that the amount of sunlight reaching the planet's surface has been going down by about 3% a decade for the last 50 years. Given the fact that we are in a warming world, this result doesn't seem to make sense, and it took a while before scientists began to work out what was behind this peculiar phenomenon. Now we know, and the results give us good reason to be even more alarmed. It turns out that we don't just emit greenhouse gases when we burn fossil fuels; we also release vast quantities of aerosols, such as sulphates that seed dense white planet-cooling clouds and hazes. It seems that the sulphates we emit can diminish rainfall over large areas, for they seed very small water droplets that are less likely to tumble out of the sky as rain. Clouds that don't produce rain have increased cooling effects simply because they stay in the sky longer. This extra cooling could act to prevent evaporation from large water bodies, as might well have happened over Lake Tanganyika, where aerosols could have decreased the amount of sunshine reaching the lake by 10%, leading to a 15% decrease in rainfall.

Overall, then, our atmospheric pollution is having two opposite effects: one warming and the other cooling, with warming the stronger of the two. The cooling effect of aerosols in the northern hemisphere has been so large that it may have changed the pattern of the African monsoon, bringing the droughts and famines that recently killed so many thousands of people in the Sahel region of North Africa. If the effect intensifies, there is a danger that the Indian monsoon could also be affected, with the potential loss of not thousands but hundreds of millions of lives. The good news is that we can easily do something about global dimming, for it is easy enough to scrub the sulphur out of our fuels and chimneystacks. We've been doing just this over the past few years, and global dimming is indeed diminishing. But there is a tragic irony, for as we reduce sulphates and other aerosols we lose the cooling effects of the clouds and hazes they seed, and the warming increases. This means that models that the IPCC used for the TAR may well be underestimating the extent of the warming to come by not incorporating the effects of dimming or its removal. It could well be that temperatures in the cleaner air will rise twice as fast as was previously thought, and that some of the critical tipping points will be surpassed as soon as 25 years from now. This would imply global warming in excess of 2^0C, at which point the Greenland ice sheet would begin its irreversible melting. By 2040 the planet could well have warmed by 4^0C, triggering the irreversible dieback of the Amazon forest. This would release even more carbon into the atmosphere by the end of this century, warming the world by 10^0C, at a pace more rapid than any other previous episode of natural warming. With this amount of warming, methane hydrates would begin to break down, releasing their vast store of methane into the atmosphere. Eventually Gaia would probably settle into a new hot state, bearable for her, but immensely dangerous and uncomfortable for us.

Perhaps we can at least be mildly grateful that another potential negative feedback has recently come to light. In a warming world, hurricanes could become far more powerful and destructive than ever before, as Katrina so grimly demonstrated. But strangely enough, the new super-hurricanes could set off a negative feedback that could take some of the heat out of climate change. Scientists analysing images from the SeaStar satellite watched in amazement as each of thirteen hurricanes crossing the Atlantic from 1998 to 2001 stirred up nutrients from the sediments below that fed phytoplankton blooms on the ocean surface for up to three weeks after each hurricane had passed. By removing carbon dioxide from the air and seeding clouds the phytoplankton blooms must have cooled the Earth, but as yet no one knows by how much. Unlike global dimming, here is a negative feedback with no strings attached, but, even if it turns out to be important, we still have no excuse for destroying Gaia's wild places or for extracting ever-increasing amounts of minerals from her already ravaged crust.

There is a remarkably solid consensus amongst climate scientists about the very real dangers of climate change. But it would be unfair to end this rapid overview without mentioning the criticisms of the small band of climate sceptics who have argued that the observed warming is due to natural variability in the activities of the sun and volcanic emissions. It is true that these important effects account for about 40% of the observed variation, but we now know that they have helped to cool rather than warm the planet during the last quarter century, and are unable to account for the 0.6°C warming trend of the last 30 years. So the sceptics have been forced to accept that warming is happening, even though most are paid, in one way or another, by the oil industry, including ExxonMobil, a major funder. The sceptics have shifted their ground from denial to arguing for the lower end predictions of the IPCC, which more or less favour business as usual. One key sceptic, the infamous Bjorn Lomborg, rightly points out that combating climate change will cost trillions of dollars, but he suggests that this huge sum of money would be better spent on dealing with AIDS and on bringing water and sanitation to the poor of the global South rather than on climate change, which to him is a minor problem. But according to Stephen Schneider, one of the world's leading climate scientists, the trillions of dollars required to tackle climate change now don't add up to much in relation to the amount of wealth generated by a growing global economy, nor to the much higher costs of dealing with the consequences in the future. He points out that acting now to avert climate change would have no significant negative effect on our future wealth.

Perhaps the sceptics should consider what happened during the Eocene period some 55 million years ago, when the Earth warmed by 5°C in the tropics and 8°C in the temperate regions due to a massive injection of greenhouse gases similar in magnitude to our own untimely gaseous exudations. Then, under a sun some 0.5% less bright than today's, the warming gases were vented either from the dissolution of methane clathrates or from the melting by an undersea volcano of a large deposit of buried organic carbon. Even though Gaia's great wild ecosystems were everywhere intact during the Eocene, it took 200,000 years for the biologically assisted weathering of granite and basalt to cool the planet. And here is the point for the sceptics to ponder—we have partially disabled Gaia by taking over about half of her land surfaces, and the sun is now hotter, so it seems likely that we are propelling our planet into an Ecocene-like 200,000-year warm period, in which most of the low and middle latitudes could become deserts. This is effectively an infinite span of time for humans—our descendants will never directly experience the wild beauty of vast tropical forests, or the miracle of the great wildebeest migrations on the plains of Africa.

And now, a note of caution. We need to reflect on what deeper meanings we might discern in the great whirling feedbacks that we have been considering. Are they, perhaps more akin to the psyche of the very Earth itself than to the so-called 'objective' processes that we describe with our sciences? If Jung is right, if our psyche is none other than the psyche of nature, then the feedbacks we have explored are as much a part of us as the churning of our intestines, as the rhythm of our breath and as the coursing of our blood through our veins. If so, then by wounding Gaia we wound ourselves, both physically and psychologically. Perhaps the most profound way we can make peace with Gaia is to feel ourselves extending outwards beyond our skins into the wide, living world of Gaia's gyring, eddying 'circles of participation'. Perhaps only then will the painstaking work of science have finally done its deeper work of bringing us home to the great living community in which we are indissolubly embedded. Otherwise, for us, our animate Earth will continue to become an increasingly desperate Earth.

Chapter 9

Gaia and Biodiversity

It was a rainy English day, and Julia, Oscar and I had gone off to Bristol for Julia's talk at the Green Food Festival about vegetarian cooking. Afterwards I went off to explore the lovely dockland area with its art galleries and trendy restaurants. Wandering into the foyer of the IMAX theatre, I saw something dramatic which caught my breath. A large metal Earth, about three metres in diameter, hung from the ceiling near the ticket office. Thick copper strands criss-crossed its hollow interior, and sheet metal continents curved gracefully over the hollow sphere. But what intrigued me most of all was that the whole thing was studded with TV screens, several for each continent and others for the oceans. Each TV showed scenes of the wild world of the region it represented. In Africa, you could see the dust clouds sent into the sunset by the great wildebeest migration; in South America, birds and butterflies moved languidly through the tropical forest; in Indonesia, orangutans swung from the trees; in Eurasia there were deer and herds of wild bison grazing in open forest glades; in Australia there were lovely scenes of kangaroos in the wild bush, and in the oceans you could see dolphins swimming in crystal blue waters and huge manta rays plunging into schools of tiny fish. I stood transfixed as the installation propelled me into a deep experience of Gaia's living biodiversity. My body tingled with a sense of the trillions of organisms teeming over the animate Earth, each contributing in its own small way to the movements of vast quantities of chemical beings in and out of the air, the rocks and the waters of our living planet. Once again, for a few brief moments I had been *Gaia'ed*. I had experienced biodiversity as a powerful force that keeps our planet alive.

Biodiversity is the diversity of life at various levels of organisation, ranging from genes, species, ecosystems, biomes and landscapes. As far as we can tell, the Earth just before the appearance of modern humans was the most biodiverse it has ever been during the 3,500 million years of life's tenure on this planet, and before we began to upset things the Earth hosted a total of somewhere between 10 and 100 million species. The fossil record shows us that there have been five mass extinctions in the last 400 million years or so, all due to natural causes such as meteorite impacts, flood basalt events, or possibly because of drastic internal reorganisations within biotic communities; but the greatest and fastest mass extinction is happening now, and is entirely due to the economic activities of modern industrial societies. We are haemorrhaging species at a rate up to 1,000 times the natural rate of extinction, or, more prosaically, every day we are losing 100 species, mostly in the great tropical forests because of our endless desire for petroleum, timber, soya, palm oil and beef. Coral reefs and the marine realm in general are not exempt from our destructive attentions either. The list of atrocities which our culture has perpetrated on the living world makes for chilling reading. According to E. O. Wilson and Paul Erlich, two of the most eminent ecologists of our day, a quarter of all the organisms on the Earth could be eliminated in 50 years. By the end of the 20th century, about 11% of all bird species, 18% of mammals, 7% of fish and 8% of all the world's plants had been threatened with extinction. According to the Living Planet Index, in the period from 1970–2000, forest species declined by 15%, fresh-water species by a staggering 54%, and marine species by 35%.

Does the current mass extinction really matter? What does biodiversity do for Gaia, and for us? To anyone who is deeply in touch with nature, it is absurd to ask these questions—clearly the current mass extinction is a crime of vast proportions. Our intuitions and deep experiences of the more-than-human world tell us that biodiversity gives us three key benefits that Aldo Leopold talked about: integrity, stability and beauty. But what does science have to say about the importance of biodiversity? To explore this question we need a systems diagram showing how biodiversity contributes to the well-being of Gaia (Figure 40).

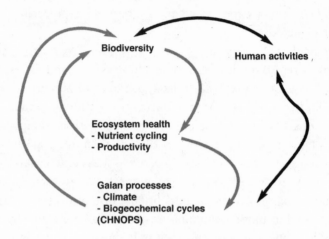

Figure 40: The importance of biodiversity for the health of Gaia
(arrows depict influences, not direct couplings).

Firstly, human influences act directly on biodiversity, or indirectly by changing Gaian processes such as climate, biogeochemical cycles and other global processes. Human-induced changes to biodiversity could then affect aspects of ecosystem health, such as how well an ecosystem resists and recovers from disturbances, how well it recycles its nutrients and how reliably, and how much biomass it produces over a given period of time. These various aspects of ecosystem health could feed back to influence biodiversity, as changes in nutrient cycling or productivity impact on the species in the ecosystem. Ecosystem health could also have big impacts on Gaian processes, such as the abundance of greenhouse gases in the atmosphere and the overall albedo of the planet, both of which influence climate. Every species has a preferred climate in which it feels most comfortable, so Gaian processes feed back to influence biodiversity. Lastly, altering biodiversity could expose human activities to feedbacks from two directions: directly from changes to biodiversity, and indirectly if ecosystem health and Gaian processes have been affected. Let's look at each of these relationships. Firstly, how are human activities influencing biodiversity? The answer has been summarised in the famous acronym 'HIPPO', which tells us that our lethal impacts on biodiversity are, in order of importance: Habitat destruction and fragmentation, Invasive species, Pollution, Population and Over-harvesting.

Habitat Destruction: Before the beginning of widespread destructive human impact during the 19th century, Gaia was clothed with a continuous cover of wild habitats that melded gently into each other according to how climates varied over her surface. If we had been standing in Britain after the last ice age was well and truly over some 10,000 years ago, we could have walked all the way from the south coast of England to the north of Scotland without ever leaving the great mosaic of wild forest and natural meadows that covered most of the country. We would have experienced a similar continuum on each and every continent. Crossing the channel to France, we could have walked all the way across Eurasia to the great rainforests of Burma, Thailand and Vietnam without ever encountering a major disturbance to nature's vast wild domain. The abundance of flying, leaping, swimming beings in this pristine state astonished the first European settlers all over the world, who quickly set about logging, hunting, fishing and clearing for agriculture with a demonic destructiveness that beggars the imagination.

Today, there is no habitat on Earth that has not been seriously degraded by humans. More than 50% of wild habitat has been destroyed in 49 out of 61 Old World tropical countries. Tropical rainforests are being cut down so quickly that by 2040 virtually no undisturbed rainforest will exist besides a few pitifully small protected fragments. All the great biomes face increasing threats, including the mangrove swamps, the wetlands, the tropical dry forests, the tundra and the boreal forests—the future for all of them looks bleak. When humans attack the great wild, they generally leave a few fragments of the original habitat here and there, perhaps out of laziness, or because of a pang of conscience, or most likely because no money could be made out of them. To begin with, these fragments are the last refuges for the great wild beings that once roamed freely over the untamed Earth, but they soon turn into death camps as the effects of fragmentation begin to bite. Each fragment is an island, often surrounded by inhospitable habitats such as agricultural land, buildings and roads that for many creatures create insurmountable barriers to foraging, dispersal and colonisation—even a small road in a nature reserve can be a daunting obstacle to tiny insects. The refugees may not be able to find the food they need in their fragments, or a mate, or even a good place to sleep. Edge effects creep into the fragments, particularly the smaller ones, making things too dry, too hot or too cold. Pests and diseases can strike down the refugees more easily in the fragments, and even if there are enough breeding individuals to keep a population going, eventually reduced colonisation from outside can lead to a seriously damaging lack of genetic diversity.

You never know who the big players are in the wild world—although they seem insignificant, the dung beetles of the Amazon are critically important for the health of the whole forest. Near Manaus, in the Amazon region of Brazil, a small dung beetle searches for food on the dry leafy floor of a small forest fragment left behind when the surrounding forest was cleared for pasture in 1982. In the old days, when the forest was entire, a whole host of dung beetle species, large and small, killed off parasites, buried seeds and ensured that precious nutrients were quickly recycled as they fed their underground larvae on buried dung. But in the forest fragment there is little dung around, for most of the monkeys and birds that provided it in abundance before the forest was fragmented died or left a long time ago. Now there are fewer kinds of dung beetle, and those that remain are smaller and not very numerous. The dung beetle extinctions happened in many ways. Hot, dry winds searing in from the pasture outside the fragment wiped out several species by killing off their larvae. For many species there just weren't enough mates to go around, and the inhospitable pasture prevented other beetles from colonising the fragment, to boost numbers and bring in new blood. The consequences for the fragment's remaining denizens have not been good. There are more diseases amongst the few birds and mammals that remain, nutrients are washed away by heavy rains before roots can capture them, and the seeds of many plants have not been able to germinate. Seemingly insignificant, the dung beetles of the Amazon are major players in their ecological community— they are one of the *keystone species* of the forest.

Introduced species: These can cause extinctions even in areas where there has been very little habitat fragmentation, and wipe out more species than pollution, population pressures and over-harvesting put together. They come from all over the world: the goats, pigs, cats, rabbits and many others, brought to places they could never have reached without the help of humans. About 4,000 exotic plant species and 2,300 exotic animal species have been brought to the United States alone, threatening 42% of species on the endangered species list and causing about $138 billion of damage every year in sectors such as forestry, agriculture and fisheries. Introduced species often do well in their new locales in the absence of their natural predators and diseases. Most don't do much damage, but a small minority take hold and do massive harm. Some are predators that exploit defenceless native prey species. A famous example is the brown tree snake (*Boiga irregularis*), a native of the Solomon Islands, New Guinea, northern and eastern Australia and eastern Indonesia. Introduced to some of the Pacific islands, it has virtually wiped out many endemic bird species. On Guam alone it is responsible for driving twelve to fourteen endemic bird species beyond the point of no return. Other introduced species are powerful competitors, like the American grey squirrel (*Sciurus carolinensis*) that seems to have pushed out the native red squirrel (*Sciurus vulgaris*) in most parts of Britain.

Pollution: Rachel Carson's seminal book, *Silent Spring,* was instrumental in starting the green movement by bringing the dangers of pesticides to our attention in 1962. Since then, pollution of many kinds has become alarmingly widespread. We are only too aware of gender-bending chemicals in water, and are well informed about atmospheric pollution such as acid rain from power stations and cancer-causing soot particles. One of the most insidious pollutants today is carbon dioxide, which paradoxically is also an essential nutrient for plants, which they harvest from the atmosphere. But it is also a greenhouse gas, and too much of it causes the climatic mayhem that leads to extinctions.

Population: This refers to the explosive growth in human numbers, especially since the industrial revolution. The current world population stands at 6.4 billion, and is projected to level off at around 10 billion by 2150. People need land, water food and shelter, and have to satisfy these needs by destroying wild nature. But it is not just a question of sheer numbers, for the amount of resources consumed by each person is what really makes a difference to our impact on the planet. Paul Erlich devised his famous I=PAT equation (pronounced IPAT) to make this point. 'I' stands for impact; 'P' stands for population, 'A' stands for affluence and 'T' stands for technology. Human impact is a product of the last three terms, so that it is possible to have a high population so long as people don't have much money to spend on industrial products. In the current economic climate, all the terms on the right hand side of the equation are increasing alarmingly. Today, the world's middle class number about 20% of the population, but they consume about 80% of the available resources. An oft-quoted fact: if everyone in the world were to consume as much as the average American, three to four extra planets would be required to provide the raw materials.

The huge pressures of the human population drive all the other causes of extinction, including the last of them all, over-harvesting.

Over-harvesting: About one third of endangered vertebrates are threatened in this way. Often the over-harvesting is carried out by poor rural people left with no other means of surviving after they have been forced off their lands by global economic forces. The rich countries of the North are also responsible for over-harvesting, and are especially responsible for driving several key fisheries to the point of extinction—the Grand Banks and the North Sea cod fisheries are sad examples. Many of the world's great whales—the right, the bowhead and the blue—had been pushed to the edge extinction by the early 20th century. Detailed mathematical models designed to calculate 'maximum sustainable yield' for some of these species were spectacular failures that led to catastrophic declines. Illegal whaling has been blamed for this, but the difficulties of observing and quantifying whale behaviour in the wild were also responsible. Many whale species have been protected to some extent since 1946, and a few, like the minke whale, are recovering, but many smaller cetaceans such as dolphins are killed every year when they become entangled in the nets of the fleets that are decimating the world's fisheries.

Vanishing Species

Sometimes one experiences the mass extinction in rather peculiar ways. It had been one of those rare English summer days when the sun is hot and bright and the air is kept cool by gentle breezes playing amongst the dappled leaves. Out in the woods the insects were humming. I walked back along the path to Schumacher College, as I had so many times, often under grey skies and in grey mood. But this time, in the bright sun, something was different. I caught sight of a blackbird flying into a tree in one of the gardens, and immediately the sense of the African bush was there. This was of course a tree in a tame English garden on a rare sunny day, but it was also, simultaneously, a wild bird-plum tree in the Okavango delta in Botswana, and the bird that had flown into it was both blackbird and red-billed hoopoe. And I, who saw the bird that was hoopoe flying into the tree that was bird-

plum also became two-fold: myself in England, and the other out in the deep wild spaces of the great bush country. My English self knew the brokenness of the English countryside with its intensive agriculture, its pubs, pylons, roads, housing estates and traffic noise. The other entered the deep conversation between wild things that know they are wild and who know the profound earthly meaning that resides in nature when it is free of the intrusive effects of our culture. This other self, along with the bird-plum and the hoopoe, knew that everything that happens in the bush is as right and as beautiful as a freshly painted sand mandala, with its power, innocence and fragility. As I walked on towards the College in the bright sun, there was a leopard sleeping in the arms of a young oak tree, and hornbills played amongst the cherry trees. Then, closer to the old buildings, a whole host of species, now slowly winking out of existence as their habitats fall to chain-saw and pollution, revealed themselves: toucans, large beetles from the African rainforests, drongos, cassowaries. The gong spoke—time for supper. The spell was broken, and I was left with a painful question: why must all these beings perish?

Biodiversity and Ecological Stability

Is it conceivable that the huge losses in biodiversity could feed back to influence the human enterprise in particular localities? To answer this question, we need to explore two further questions. Do organisms living in a specific place—in one of the biomes represented by the TV images in the metal globe for example—link up into an ecological 'superorganism', with valuable emergent properties such as climate regulation, better water retention, nutrient cycling and resistance to diseases; or are they are no more than collections of individually selfish organisms, each out to exploit as many of the available resources as possible, even to the detriment of the ecological community that enfolds them? If the former is true, then we will need to protect entire ecological communities in order to preserve the ecosystem services they provide. If the latter is the case, then we need only bother to look after the key players, or to introduce those of our own choosing.

These questions occupied the minds of the founding fathers of ecology in the first half of the 20th century. The American ecologist Frederick Clements, one the most influential ecologists of his day, studied how plants colonise bare ground. He noticed that there was a series of stages, beginning with an inherently unstable plant community and ending up in a stable climax community in balance with its environment. In Devon, from where I write, bare ground is first colonised by annual herbaceous plants, then by brambles and shrubs, and eventually by oak forest, which grows here because the mix of soil, temperature, rainfall and wind are just right. For Clements, the development of vegetation resembled the growth process of an individual living being, and each plant was like an individual cell in our own bodies. He thought of the climax community as a *complex organism* in which the member species work together to create an emergent self-regulating network, in which the whole is greater than the sum of the parts.

Within the scientific community, a struggle ensued between the organismic views of Clements and the objectivist approach of the Oxford botanist Sir Arthur Tansley. Tansley declared that plant communities couldn't be superorganisms because they are nothing more than random assemblages of species with no emergent properties. He found Clements' views difficult to accept because they challenged our legitimacy as humans to remake nature as we liked. Tansley wanted to remove the word 'community' from the ecologist's vocabulary because he believed, in the words of Donald Worster that "There can be no psychic bond between animals and plants in a locality. They can have no true social order." Tansley represented a breed of ecologists who wanted to develop a completely mechanistic understanding of nature, in which, according to Worster, nature is seen as "a well-regulated assembly line, as nothing more than a reflection of the modern corporate state". For Tansley, agricultural fields were no better or worse than wild plant communities. To paraphrase Worster, the reduction of nature to easily quantified components removed any emotional impediments to its unrestrained exploitation. Ecology, he says, took on the economic language of cost-benefit analysis, but economics learned nothing from ecology.

Which approach best describes biotic communities: organism, or mechanism? Out in the flatlands of Minnesota, at a place called Cedar Creek, a long-term experiment is in progress that could have a bearing on these questions. A strange chequerboard of metre-square plots filled with prairie plants dots the landscape, tended by David Tilman, one of the world's leading ecologists, who has spent years investigating the relationship between biodiversity in his plots and the ability of the small ecological communities they contain to produce more biomass by capturing sunlight and to survive stress. Tilman and his numerous assistants have set up hundreds of plots, each with a different number of species chosen from the native flora of the immediate locality. Half-way through one of these experiments, Minnesota experienced a severe drought, and to Tilman's amazement the plots that survived best were those with the highest biodiversity. This was evidence in favour of Clements and the organismic view, for the most diverse plots seemed to have developed a powerful emergent protective network as their various members melded their individual survival skills into a greater whole. But there were critics. They pointed out that because Tilman had fertilised his plots with different amounts of nitrogen, the differences in drought resistance were due to this, and not to the effects of species diversity.

To eliminate this possibility, Tilman established a more extensive experiment in 1994, using 489 plots of two sizes with different amounts of plant biodiversity, seeded in identical soil and chosen from a maximum of four 'functional groups', each with different survival strategies: broad-leaved perennial herbs, nitrogen-fixing legumes, warm season grasses and cool season grasses. This time, the more diverse plots produced more biomass, fixed more nitrogen, were better at resisting weed invasions and were less prone to fungal infections. The best plots were those that hosted a variety of species from each of the four functional groups. Once again, here is evidence that diverse biotic communities resemble organisms with powerful emergent properties. But the news was not all good, because Tilman found that the benefits of having extra species in the community peaked at around five to ten species. Beyond that, extra species didn't seem to improve ecological performance—what mattered most was having at least one member of each functional group. Some ecologists say that these results show that most species in wild ecosystems are dispensable, and that the extinction crisis gives us nothing to worry about. But how are we to know which species are expendable and which aren't? Since we can't tell which are the keystone species, it makes more sense to protect as many species as we can. Furthermore, there is almost certainly an 'insurance effect' at work, in that more biodiverse communities are more likely to contain species that can take over the jobs left vacant by any keystone species that have disappeared.

Tilman's approach has recently been extended by the BIODEPTH project, in which plots with different amounts of native grassland biodiversity were set up in eight European countries, from the cold north to the warm south. Despite the wide range of climatic conditions, high biodiversity in each country was strongly correlated with improvements in many key ecological functions such as nutrient cycling, resistance to predators and biomass production—once again, good evidence in favour of the organismic view (Figure 41).

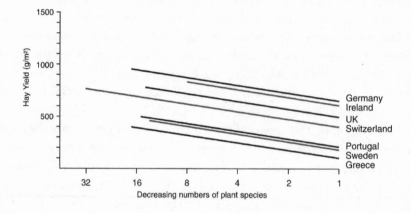

Figure 41: A key result from the BIODEPTH experiment. Each line is a statistically significant fit through the raw data points for a given country (Hector 1999).

Laboratory experiments also tend to support the idea that biodiversity improves the health of ecosystems. Scientists at Imperial College, London, have developed the 'Ecotron', a series of chambers with controlled light, temperature and humidity levels which house artificially assembled ecological communities, each with differing amounts of biodiversity. The main result of this research is that more diverse communities fixed more carbon dioxide from the air. This may seem to be a fairly mundane finding, but it caused a stir in scientific circles as it showed that biodiversity could have a key role to play in absorbing some of the vast amounts of the Earth-warming carbon dioxide gas that our economy is emitting into the atmosphere. In other words, terrestrial biodiversity may be of major use to us in helping to combat global warming, at least in the short term. New work in the Ecotron mimicked the elevated carbon dioxide and temperature that are expected with climate change. The surprising result was that climate change had little impact on the fauna and flora living above ground, but the community of soil organisms was greatly altered. More carbon dioxide in the atmosphere stimulated photosynthesis amongst the plants, which then transported some of this carbon to their roots as sugars. The extra soil carbon changed the community of soil fungi, which in turn changed the community of fungus-eating springtails. These changes in below ground ecology could, if writ large, have a massive impact on nutrient feedbacks and carbon storage in soils, but as yet no one knows whether this means that soils will be able to hold more or less carbon. The fact that there was a change is worrying, and could have an effect on future strategies for dealing with climate change.

In another series of experiments, scientists created artificial ecological communities by seeding glass bottles containing water and nutrients with differing diverse communities of bacteria and their larger protist predators. In these experiments, greater diversity led to less variability in the flow of carbon dioxide in and out of the community. The message here is that more diverse real-world communities could provide more predictable and dependable emergent ecological functions, such as carbon capture and storage.

Recent mathematical modelling has also contributed to the new understanding of the relationship between biodiversity and ecological health. We now know from detailed fieldwork that ecological communities are replete with weak interactions, with many predators focusing on eating a few individuals from a fairly wide range of species. Models that take account of these insights show that virtual communities with realistic feeding relationships and abundant weak interactions are more stable than previously thought possible. Another group of mathematical models known as 'community assembly models' work by creating a pool of virtual plants, herbivores and carnivores, each with its own body size and preferences for food and space. One species at time is placed in an 'arena' where it interacts with other species that are already present. After a while, an astonishing thing happens—persistent communities self-assemble with a final membership of about 15 species. As the number of species builds up, it becomes harder and harder for an invader to find a toehold in the nexus of interacting species. Communities that have existed for longer are harder to invade than newly established ones, strongly suggesting that communities develop an emergent protective network that becomes more effective as the community matures. Amazingly, the challenge for an invader lies with the community as whole. In a mature, well-connected community, an inferior competitor has a better chance of surviving an invasion from a superior competitor than it would as a member of a less well connected community.

It is astonishing that these sorts of true-to-life emergent properties should appear in mere mathematical communities. All the research we've looked at so far, from field, laboratory and computer modelling, tends to support Clements' idea that ecological communities can indeed be thought of a superorganisms which function more smoothly and predictably as their biodiversity increases. But perhaps Clements and Tansley were both right after all; perhaps each had seen different sides of the same coin. If so, there is nothing inevitable about which species will colonise a bare patch of land, or indeed nothing inevitable about how a particular succession will progress (Tansley), but as soon as the species in a given place begin to web themselves together, the whole community becomes a superorganism with powerful emergent properties (Clements).

This implies that the plant community is an animate entity which experiences its surroundings as a single being capable of responding to threats and dangers somewhat like our own sensing bodies. In Clements' ideas, as in those of Lovelock, there is a strong animistic or panpsychist undercurrent that has much in common with the ideas of the highly influential American thinker Gregory Bateson (1904–1980), for whom the whole of nature is a vast interconnected "mind" that exists by virtue of the information flows between all of its components, even those that are not alive in the biological sense. For Bateson, an ecological community is a coherent being with its own emergent "mental" state that arises out of the sum total of all its interactions. He pointed out that our human proclivity for rational thought interferes with our ability to intuitively perceive the "pattern that connects"—a deeper reality that we can only reach through the appreciation of *beauty*. Bateson's way of seeing the world echoes that of Aldo Leopold, for whom beauty was a guide to right relationship with nature. If this is correct, then in order to fully understand (rather than explain) the profoundly animate quality of Gaia's ecological communities, we need to complement reason with the knowledge given us by our sensing bodies, by our appreciation of values and by our intuitions. Only then are we capable of perceiving the beauty of the living world around us.

Biodiversity Symphony

Go outside into your nearest area of relatively undisturbed nature. Spend some time taking in the sounds, sights, touches and smells around you. Regard them all as communication from the animate Earth that surrounds and enfolds you. Relax deeply as you absorb these messages.

Now walk about slowly and become aware of the different species around you, be they plants, animals, fungi, algae or microbes. Never mind their names, or the details of their biology, just spend time sensing each being as an animate presence, as a living creature engaged in the self-display of its own particular innate qualities. Does it make any sounds? What colours, shapes and textures is it presenting to you? What emotions does it evoke in you? Regard each species as a living personality; consider your interaction as a two-way communication between two sentient beings. Do this for as many species as you can, for as long as possible.

When you have absorbed as much detail as you can, find a place (preferably on the ground) to sit or lie down. Close your eyes and take a few deep breaths. Allow all the impressions you have gathered from all the beings you have met to meld into an emergent appreciation of the life of the ecological community as a whole. Feel the invisible connections that tie the species that you've encountered into a coherent oneness. What does the quality of the whole 'taste' like? Is it dry and spiky, wet and windswept, lush and gloomy? What is it saying to you? If the qualities form themselves into words, write them down, letting them spill onto the page. Sketch or paint any images that might arise, hum any tunes.

Now get up and walk around again, and repeat the process two or three times, going from sensing the individual beings in the ecological community to intuiting its emergent quality, its participatory intelligence, its gesture.

Finally, lie or sit on the ground with your eyes closed. Now, starting from your immediate ecological community, slowly expand your awareness to take in Gaia's great ecological communities: rain-forests, tundra, boreal forests, temperate forests, heathlands, deserts and high mountain regions, the oceans. Even though you have never met most of the more-than-human beings that live in these won-drous domains, you are able to connect with a sense of the stunning biodiversity they contain.

Feel all Gaia's living beings crawling, swimming, flying, walking, growing and running over her Earthly surface, bringing the rocks to life, shaping the very taste of the air itself, its temperature, pressure and humidity. Feel the rocks, the air and the water giving themselves freely to the exciting adventures that Gaia's diverse living beings invite them to participate in.

Biodiversity and Climate

So far we've looked at the effects of biodiversity on ecological health at the local level, but could there be a relationship between biodiversity and the health of the planet as a whole? This question, considered absurd by the scientific community as recently as ten years ago, is now beginning to loom large in the minds of scientists trying to understand how humans are changing the Earth, which they now recognise is a fully integrated system in which life is a key player.

It is now generally agreed that life affects climate in at least two major ways: by altering the composition of the atmosphere, and by changing how solar energy heats up the Earth's surface and how this heat is distrib-uted around the planet. But how could biodiversity be involved in making these globally important processes work more effectively? The Ecotron and BIODEPTH experiments have taught us that diverse ecological com-munities on the land can change the composition of our atmosphere by increasing the absorption of carbon dioxide. It is almost certain that bio-diversity in the oceans also enhances this effect. Marine phytoplankton use carbon dioxide for photosynthesis much as land plants do, drawing it out of the air and into their tiny bodies. Dead phytoplankton sink, taking car-bon that was once in the atmosphere with them to a muddy grave in the sediments below. This is the biological pump that we encountered earlier, and it is almost certainly more effective at removing carbon dioxide from the atmosphere wherever there is greater marine biodiversity, partly because the larger phytoplankton in more diverse communities increase the slow drift of carbon to the ocean depths.

Biodiversity may also influence the absorption and distribution of energy from the sun. It could be that more diverse communities on land and in the ocean are better at seeding clouds, but this remains to be seen. What is more certain is that a greater diversity of land plants could enhance cloud-making and energy distribution in two other important ways: by transpiring more water from the soil through roots and out into the air from pores on the undersides of leaves, and by providing more leaf surfaces from which rainwater can evaporate directly.

A big rain storm has just finished watering several hundred square kilometres of Amazon forest. The leaves are all wet, and those at the top of the canopy glisten in the early afternoon sun. Some of the energy in the sunlight passes deep into the leaves where it fuels photosynthesis, but a fairly large portion is absorbed directly by the recently arrived film of water on the leaf surfaces. As the water molecules receive their gift of solar energy they begin to gyrate like inspired dancers, and when sufficiently energised they dance their way into the air as water vapour. This is evaporation. In the case of a leaf drying in the sun, solar energy which might have heated the leaf is transferred to water vapour, and as this is swept away by the wind, the leaf is kept cool, just as we are when we sweat.

The energy held in water vapour can be released as heat whenever condensation converts it back into liquid water. This energy is called 'latent heat' because it remains 'invisible' until condensation happens. On the other hand, any solar energy absorbed by the surface of the leaf causes the molecules there to vibrate and to immediately re-emit the energy as sensible heat, which you can detect directly with your skin or indirectly if you have an infrared sensor.

But it is not just rainwater that evaporates from the surface of a leaf; so does water that has travelled from the soil into the plant through tubes leading all the way from the roots to the thousands of microscopic pores beneath a leaf's surface. This water, carrying with it life-giving nutrients from the soil, eventually passes through the leaf pores into the air, a process known as transpiration. Amazingly, plants keep the flow of water going without the kind of muscular contraction seen in animal circulatory systems. They do this by continually and deliberately leaking water through the pores, thereby creating a mysterious kind of 'suction' that draws in new water all the way down at the roots. On warm days, water entering a leaf from the soil is heated up by the sun's rays, and passes out of the leaf pores as water vapour. The summed effect of evaporation of water from leaf surfaces and transpiration of water from within the plant is considered to be a single process known as evapotranspiration, which is vitally important for Gaia's climate. Because of it, a huge amount of solar energy is stored as latent heat in water vapour that can travel long distances before condensing to release its energy as heat, sometimes thousands of kilometres away. But evapotranspiration also has local effects. In the deciduous forests of the northeastern United States, temperature rises steeply in the early spring when, unimpeded by leaves, the sun's rays warm the ground. But as the leaves unfurl and swell out to their full size, the rate of temperature increase drops dramatically because evapotranspiration cools and moistens the air.

Foliage is thus very important in regulating the surface climate. In general, the more leafy a forest, the more evapotranspiration and so the more cloud production, local rainfall, local cooling and plant matter production by photosynthesis. A more diverse flora almost certainly improves transpiration by providing a bigger and more varied mat of below-ground root structures with better water-trapping abilities, and it could also enhance evaporation by providing a larger and more complex total leaf surface area from which rainwater can evaporate. Both of these effects send more water vapour into the air for cloud-making. Some plants evapotranspire more than others. Because they have far fewer leaf pores, needleleaf trees pass less water into the air than their broadleaved cousins, and as a consequence needle leaves seed fewer clouds, thereby keeping themselves warmer—an advantage in the high latitudes.

Another climatically important characteristic of vegetation is its roughness, a measure of how much resistance plants give to the wind. When wind blowing over the land surface encounters plants such as trees, grasses and shrubs it transfers some of its energy to the leaves, making them dance about. This sometimes frenzied leafy dance mixes the air, making both evapotranspiration and the transfer of sensible heat from leaf to air much more effective than on a perfectly still day. The higher up the canopy you go, the more efficient are these transfers of energy from wind and sun to leaf. A dense rainforest canopy, with its high roughness, will transfer much more energy to the air than the far less leafy, low roughness grasses in a savannah. The intricate leaf surfaces of a more diverse flora create a rougher land surface that increases air turbulence, and this could well increase the transfers of heat and moisture to the air, influencing weather patterns on both local and global scales.

These impacts of biodiversity on local and global climates in turn feed back to influence biodiversity itself. Clouds seeded by the Amazon forests keep the forest cool and recycle its water, thereby allowing the forest to persist and preventing the encroachment of the nearby drought-tolerant savannah. The heat released when the clouds condense helps to configure the Earth's climate system as a whole into a state that favours forest growth in the Amazon region. Herein lies a great lesson for living in peace with Gaia: the very structure of an ecosystem—namely which species are present, the depths of its roots, the extent of its leafiness, its albedo and its release of cloud-seeding chemicals to the air—all have massive effects not only on climate both locally and globally, but also on the great cycling of chemical beings around the planet.

We have seen how biodiversity is a key player in creating habitable conditions on the Earth, including a climate that favours our own existence. Biodiversity also provides us with a host of other benefits, such the stabilisation of soil, recycling of nutrients, water purification and pollination. These benefits have been called 'ecosystem services' by a new breed of economists who are attempting to calculate how much these services are worth in financial terms. The results are staggering—in 1997 global ecosystem services were worth almost twice the global GDP. Recently, the results of the most comprehensive survey of the state of the world's ecosystem services were made public. The Millennium Ecosystem Assessment, compiled by 1,360 scientists from 95 countries, deliberately took the approach of looking for the interconnections between human well-being and ecosystem health. The results make sobering reading—in all, 60% of the ecosystem services investigated have been degraded. Human activity has changed ecosystems more rapidly in the past 50 years than at any other time in human history. About 24% of the planet's land surface is now under cultivation; a quarter of all fish stocks are over-harvested; 35% of the world's mangroves and 20% of its coral reefs have been destroyed since 1980; 40%–60% of all available freshwater is now being diverted for human use; forest has been completely cleared from 25 countries and forest cover has been reduced by 90% in another 29 countries; more wild

land has been ploughed up since 1945 than during the 18th and 19th centuries put together; demands on fisheries and freshwater already outstrip
supply; and fertiliser runoff is disturbing aquatic ecosystem services. The
report makes it abundantly clear that the UN's Millennium Development
Goals of halving poverty, hunger and child mortality by 2015 cannot be
met unless ecosystem services are nurtured and protected, because it is the
poor who are most directly dependent on these services, particularly for
fresh water and protein from wild fish and game. Furthermore, it has
become abundantly clear from a handful of successful projects that the
way forward lies with encouraging local people to become involved in
protecting their own ecosystem services. This has worked well in Fiji,
where local fishermen established restricted areas that reversed serious
declines in fish stocks, and in Tanzania where villagers now harvest food
and fuel from 3,500 square kilometres of degraded land that they were
allowed to reforest.

All of this should be enough to convince the most hard-headed
amongst us that it is very much in our own interest to maintain as much
of our planet's native biodiversity as possible; but these utilitarian arguments for protecting biodiversity may not prevent it from being seriously
degraded, for ultimately, in the words of Stephen Jay Gould, we may not
be able to save what we do not love. If we are ever to develop a world-
view that has any chance of bringing about genuine ecological sustainability, we will need to move away from valuing everything around us only in
terms of what we can get out of it, recognising instead that all life has
intrinsic value irrespective of its use to us. Scientific and economic arguments such as those we have been exploring for protecting biodiversity can
help a great deal, but on their own they are not enough. We need, as a
matter of the utmost urgency, to recover the ancient view of Gaia as a fully
integrated, living being consisting of all her life-forms, air, rocks, oceans,
lakes and rivers, if we are ever to halt the latest, and possibly greatest,
mass extinction.

Chapter 10

In Service to Gaia

Why Does Nature Matter?

Many people who come to Schumacher College ask why it matters that we are destroying so much of nature and possibly our own civilization as well. They ask: isn't what we are doing totally natural, isn't it an inevitable consequence of our deeply flawed human nature? Aren't we Gaia's way of bringing about a new mass extinction? Are we any different from any other global catastrophe like a meteorite impact or a flood basalt event? Shouldn't we accept our fate as the nemesis of the current configuration of Gaia and enjoy ourselves whilst there is time? What is the meaning of it all anyway?

These questions are disturbing because they point to a deep sense of disillusionment, meaninglessness and inevitability that trouble the increasing numbers of people who are waking up to the horror of what we are doing to the Earth. Answering 'yes' to all but the last question (which requires more than a one-word answer) is deeply discouraging and disempowering, and leads to a numbing apathy that helps no one.

The curious thing is that it is not possible to explore these questions from a purely rational perspective, for a satisfying response requires us to connect with ways of knowing beyond our thinking minds, which, after all, serve us best when they are dealing with facts, but not with sensations, feelings or intuitions. If you love art, imagine how you would feel if one day you woke up, turned on the radio and heard that all the great master-pieces in all the world's great art galleries had been simultaneously slashed and burnt by an international gang of demented fanatics, for no particu-lar reason. All the Monets, Renoirs, Caravaggios, Da Vincis, Rembrandts, gone, never to be seen again as vibrant, original canvasses. You would feel outrage, loss, sadness, and distaste, and would not take kindly to a well-meaning rationalist telling you that in the end it doesn't really matter because there will be plenty of great artists in the future to create magnif-icent new works of art. We might be able to accept this argument on a the-oretical level, but our intuitions and feelings tell us, somewhere in our guts and hearts, that each masterpiece is worthy of respect simply because it exists, and that its destruction is a heinous crime.

Similarly, when we are out in nature seeing the more-than-human world around us in all its beauty, we know intuitively that each biological species is a unique masterpiece as worthy of wonder and respect as is the work of any human artistic genius. After all is said and done, the same great creative forces of the universe have made human artists and non-human species alike. When we are in the place of awe and wonder, we know with unshake-able inner certainty that the destruction of biodiversity is a crime.

Here is another example. Any sane person knows that murder is wrong, but a rationalist could justify the act by saying that it doesn't really matter because a new person will soon be born to replace the one that has been dispatched. But murder is wrong because we know intuitively that each person has intrinsic value, and because we feel that each human being must be nurtured as a unique event in the unfolding of human con-sciousness and experience. Since there is no fundamental divide between humans and the rest of the cosmos, it is wrong, according to this deep intuition, to murder any aspect of the more-than-human world, be it a species, a river, a biotic community or a great biome, because all are imbued with intrinsic value and so are worthy of deep respect.

These two arguments work by giving the more-than-human world a sta-tus equal to that of human beings and to the works of the best human cre-ative genius. I cannot prove the equivalence; my intuition and my deep experience of awe and wonder tell me that this is the case. Unless you have had similar insights, no amount of rational argument will convince you, for we are not talking about utility, but about sanctity. This is difficult for our culture to accept, for we tend to think of anything non-human as somehow inferior, as subtly different to us, as somehow not really *alive*, as in the end as no better than a mere machine. We will not be able to understand the insight without having a certain kind of experience which no end of rational discourse can produce, the essence of which is that every speck of matter is sacred simply because it exists. If you can't accept this, then please go out into your nearest tract of free nature. If this is far away, then do all you can to make some time to get out there, hop on a bus or train, or drive if you have no other alternative. Spend as long as you can there. Sit quietly in the woods, or near a river or under the great wide sky. Let them awaken your innate capacity for direct connection with our fabulous animate Earth.

But isn't what we are doing completely natural and inevitable? Isn't it part of the tragedy of our species that we can recognise the intrinsic value of the more-than-human world whilst destroying it at the same time? Everything we humans produce is natural, including atomic bombs, GM crops, plastics, pesticides and deforestation. What makes all the difference is our level of consciousness when devising and using technology. If we are driven by greed, naked ambition, hatred and selfishness, then the outcomes are bound to be negative; but if we act from a sense of solidarity with all beings, from a desire to be of service to the great wide world, and from the deep realisation that there is only one self, which is the great Self of the universe, then we are more likely to act with wisdom and restraint. Humans are different from a flood basalt or a meteorite impact because, unlike them, we have the ability to *choose* how to be present to the world. A meteorite on a collision course with the Earth cannot suddenly change its mind and swerve harmlessly away; it has to collide with us, if the forces acting on it have configured it to do so—the meteorite has only one degree of freedom. We humans, on the other hand, are blessed with many degrees of freedom. Now that our best science has informed us of the huge ecological and social crises we are unleashing upon the world, we can choose whether to remain in the narrow, objectivist mode of consciousness that has contributed to the crises, or to act from a deeper, wider mode of consciousness in which we experience our unity with the whole of Gaia and hence understand the importance of radically changing our way of being in the world.

In this mode of awareness, we come to realise that Gaia is beyond our control—that it is impossible for us to ever be the masters or stewards of the Earth. We also come to understand that for Gaia, every living being represents a uniquely valuable mode of sentience—that it is hubris to think that we are the only sentient creatures inhabiting Gaia's ancient crumpled surface. Slowly, as we develop sensitivity to many other styles of non-human sentience every bit as important as our own, we realise that we owe our very existence to the complex planetary intelligence that has run our world without our input for the last 3,500 million years.

Another question that is often asked is this: Why does it matter about the extinction crisis and climate change? If Gaia is a great self-regulating being, won't she take care of herself in the long run? Given enough time, perhaps five million years, Gaia will recover from our onslaughts, and will once again produce a great flourishing of biodiversity as she has done after previous mass extinctions. So why worry? All of this may be true, but of course the fact that biodiversity has recovered after previous mass extinctions is no guarantee that the same thing will happen again. After all, Gaia is now older and more stressed by the sun, and the current mass extinction is happening much more quickly than any other. Nevertheless, there is at least a good chance that Gaia will eventually recover, but the question reveals an attitude which is somewhat troublesome and unhelpful, for in order to comply with it you have to mentally abstract yourself from the tragedy of what is happening and pretend that you are not fully *embedded* in Gaia as one of her humble living, breathing, animate creatures. You have to pretend that you are somehow immune to what is happening, that you can take an aloof, God's eye view of the situation. But whether you like it or not, you are utterly part of Gaia, biologically, psychologically and spiritually. Our very bodies, our dreams, our creativity, our imagination all come from her, and in end the matter that we are made of will return to her when our lives are done. Once you allow yourself to feel this deep belonging to Gaia, there is no question that what we are doing to her now is wrong, and that we have to do something about it.

So being of service to Gaia requires us to develop a deep sense of embeddedness in the life of the great planetary being that has given birth to us and to every other creature that has ever oozed, crawled or sent its roots into our planet's soil. We need to sense that our every step is taken not *on*, but *in* the Earth; that we walk, talk and live our whole lives inside a great planetary being that is continuously nourishing us physically with her miraculous mantle of green and her luscious swirling atmosphere, a being that soothes our psyches with her subtle language of wind and rain, with the swoop of wild birds and with the majesty of her mountains. We need to develop a sense of ourselves as beings in symbiotic relationship with Gaia, just as the mitochondria live in an intimate relationship with their larger, unseen host. We need to remember that our very breathing is to drink our mother's milk—the air—made for us by countless microbial brothers and sisters in the sea and soil, and by the plant beings with whom we share the great land surfaces of our mother's lustrous sphere. We need to develop a sense that Gaia really is alive, not in some metaphorical sense, but really, actually, palpably, to the extent that that you recognize the joy of sunlight on the great bare branches of winter trees as not just your joy, but as the joy of the entire cosmos revelling in sheer astonishment that such beauty could have unfurled out of itself like a young leaf in spring into the plenum of being. Let Gaia take you over—let yourself be *Gaia'ed* over and over again.

We have delved into the science of Gaia, and have seen that it gives us the best possible cognitive basis for knowing that the Earth is alive—not 'life-like'—but really alive. Can we let the science be like a juicy bone tossed to the rational mind to keep it happily chewing whilst the real work of developing our belonging to Gaia happens through our senses, our feelings and perhaps most importantly, our intuition? Let these be the gateways into our new sense of belonging in a living world, and let our reason take its rightful place as the servant of this deeper, more intoxicating knowledge. As the American poet Robinson Jeffers once said, let us "fall in love outwards", and may this falling in love spread like a contagion as far and wide from mind to mind and from sensing body to sensing body as fast as possible, for time is running out for us and for the great wild planet into which we were born, and without which our lives would hardly be worth living.

Gaia is a being so large that we can never physically see the whole of her as we can the whole of an orange, or a tree, a flower, or another human being. These are beings that we can walk round or turn in our hands, but we can't do this with our planet, at least not without the help of a lot of highly technical instruments. So how are we to develop a sense of Gaia as an entirety, as a whole being? One of philosopher Henri Bortoft's great insights is that every part of a phenomenon contains the whole, that reality is holographic rather than fragmented. This gives us a tremendously important clue about how we can cultivate our sense of belonging to Gaia: we can do this by developing a deep *love of place*. The soul of a place, when entered into with the deep interest and concern that love entails, contains the quality of Gaia as a whole being. But will any place do? What about a noisy road in the middle of a busy city? Ultimately, yes, it is possible to find Gaia even in such places. I have two remarkable friends who lived for many years near Heathrow airport, right under the flight path. From their living room window they could see the planes banking towards them in preparation for landing. The roaring and the shaking as the planes flew over them were intense and deeply stressful. My friends had nowhere else to go, so in the end they decided to surrender and slowly they learnt how to love the loud noises, the shaking, and the planes themselves. As they watched the planes coming in they would lovingly think of the wonderful places where a given plane might have come from—perhaps the coral-studded Caribbean, or the snow-covered Alps, or the sun-soaked Mediterranean. In the end, they transformed their negative feelings about their place into a sense of connection with the whole of life. Admittedly, this is a very high level of practice, which most of us would almost certainly be incapable of, but it shows what is possible.

We need to find a place we find beautiful: perhaps a park or a small garden if we live in the city, or a wood or the seashore if we are lucky enough to live in the country. Once we have found our place, we need to spend time in its silence, getting to know it intimately with our intuition and feeling, whilst using our rational mind to find out about its geology, botany and zoology, and about how humans have interacted with it over the years. We need to give ourselves time to experience the *soul* of the place, and through it the soul of the world, the *anima mundi*. A next step is to extend our love of place outwards to our local bioregion, which is, according to the Convention on Biological Diversity, "a territory defined by a combination of biological, social, and geographic criteria, rather than geopolitical considerations; generally, a system of related, interconnected ecosystems". In other words, your local watershed, your local valley, or your local tract of wild bush, jungle or desert, should you be lucky enough to live in such a glorious place. Through love of place we deepen our love of Gaia.

Love of place becomes a political act when we share it with other people. It nourishes a sense of the local human community embedded in the air, soil, water and more-than-human beings of the immediate surroundings; it nourishes local economies, and the growing of local organic food; it nourishes the raising of children in the love of place; and it nourishes art, music, literature and science, all based on what is local. In these ways, love of place is the ultimate act of non-violent resistance to the major force that is destroying the animate Earth that we evolved into. And what is this destructive force? Two words sum it up: Economic Growth.

The Trouble with Growth

Let me make it clear that by 'economic growth', I here mean growth in the throughput of physical matter into and out of the economy, not growth of non-material things such as music, ideas, information and so on, as long as the growth of these things in no way depends on increasing amounts of material throughput. Mainstream economics is obsessed with growth of the material kind. The basic idea is that to be healthy, economies must constantly increase the amounts of raw materials that flow through them in order to generate ever greater wealth, and that in order to be happy, people must have more and more of this wealth so as to have access to consumer goods. The raw materials for these goods must obviously come from nature, which economists perversely think of as an infinite repository of oil, minerals, timber, fish, and a whole host of other so-called 'resources'. But growth has failed to make us any happier, and is degrading the more-than-human world on which we utterly depend.

The payment of interest is one of the key drivers behind the growth imperative. As the economist Richard Douthwaite explains in his classic book *The Growth Illusion*, businesses need to borrow money to fund their operations, and the main way in which they pay the interest on their loans is by investing in other businesses that will yield good returns. If the economy as a whole is growing, then profits will increase across the board, and so each business will have sufficient profit from its investments to cover its interest payments. This is a classical positive feedback loop—a genuinely vicious cycle that inevitably leads to ever more growth and the social and ecological breakdown that this entails.

The major commodity that currently fuels growth though international trade is money. Each day, 1.3 trillion US dollars are traded on the international money markets in what amounts to gambling on a massive scale. The money market creates severe instability globally because huge profits can be made at a touch of a button by moving money out of one country and into another at the whim of a handful of mega-wealthy transnational corporations. Once money has been moved out of a country, its economy is vulnerable to collapse, as was the case recently in various countries around the world. The motivation for making all this money is of course the need to pay off interest and make surplus profit.

The mainstream indicator of growth is the notorious GDP—the Gross Domestic Product, which is a measure of the total value of the financial transactions that have taken place in a society over a specified period of time. If GDP goes up, then we have growth and all is well; if it goes down, there is depression and national soul-searching about what could possibly have gone wrong, as is happening now in Japan. But GDP is a woefully inadequate measure of what really matters, namely human and ecological well-being. GDP goes up whenever a financial transaction of any kind takes place, so it is extremely good at including things we normally consider to be highly undesirable, such as the cost of treating cancer and the financial implications of car accidents. It also ignores so called 'externalities', like pollution and ecological destruction—the unwanted side effects of growth. GDP around the world has grown since the end of the Second World War, but there is ample evidence that growth is making us less and less happy. In America, for example, GDP has been going up steadily since the late 1940s, but alternative indicators of human well-being, such the GPI index (the Genuine Progress Indicator—Figure 42), which take into account things that make us happy and discount those that don't, peaked around the 1970s and have remained stationary ever since, despite continued GDP growth. Clearly, growth is failing to make us happier. In the words of the Australian economist Clive Hamilton, growth has become "an inanimate object worshipped for its apparent magical powers"—it is a "fetish" that we worship irrationally because our sense of self-worth is tied up with our power to consume.

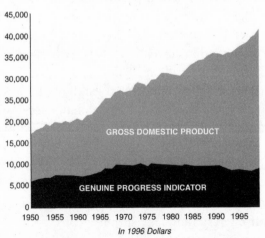

Figure 42: GDP and GPI for the United States.
(*source: www.redefiningprogress.org*)

Growth is bad for Gaia, and the reason is blindingly obvious. We live in a world of limited 'resources', which will eventually run out as the growth economy exploits them to exhaustion. It has always amazed me how economists seem unable to understand what is patently obvious to little children—that the contents of the biscuit box will run out if you eat them faster than your mum or dad can replenish them. A growing economy must eventually wipe out all the forests, exhaust all the fisheries, mine all the minerals and extract all the oil. Furthermore, growth will erode both a favourable climate and human health when Gaia's capacity to deal with the increasing volumes of waste and pollution becomes saturated.

Growth also erodes the social fabric because it sets us all against each other as we scramble to earn more and climb higher up the social ladder— it sets the rich North against the poor South, and perpetrates a new form of colonialism every bit as cruel and unkind as its more blatant manifestations in the last century and the one before.

The instruments of this new kind of growth-obsessed colonialism are the World Trade Organisation (WTO), the World Bank, and the International Monetary Fund (IMF), all of which operate completely outside of the democratic process, with unelected representatives meeting behind closed doors, accountable to no one other than their corporate masters. The aim of these institutions is to promote as much economic growth as possible by championing 'free trade', which means the free flow of goods and investment across national boundaries without let or hindrance on the part of sovereign national governments. The paramount achievements of these arrangements have been increased social and ecological breakdown. For example, trade liberalisation under WTO rules has seen to it that food grown in the North with the help of heavy government subsidies floods into food markets in the South, undercutting the price of local produce and driving thousands of farmers off their lands and into burgeoning city slums. Free trade rules also allow transnational corporations to move factories to countries where labour is cheap and taxes and environmental safeguards are minimal. Local companies have to pay taxes from which the transnationals are exempted because they can threaten to move their operations elsewhere. Governments lose revenue as a result, and have fewer resources for social welfare programmes and for environmental protection. These arrangements often have severe ecological and social consequences. The World Bank has given loans for establishing shrimp farms all over the tropics, and the money has been used to clear huge areas of mangrove swamps and wetlands for these enterprises, which sell their produce to distant luxury food markets in the West. But mangrove swamps, as we discovered during the recent Asian tsunami disaster, are vitally important for coastal protection. They also provide homes to a whole host of living beings, and of course, from a deep ecology perspective, are valuable in themselves simply because they exist.

Technological optimists tell us that we need not worry about the ecological impacts of growth because improvements in efficiency will sort out the problem. They believe that we can 'dematerialise' our industrial products so much that we will require virtually no physical matter to make them. They speak of 'factor four' and 'factor 1000', meaning that we can make the same commodities with one quarter or one thousandth of the raw materials we use today, and they rightly advocate the recycling of every last bit of matter so that industry follows nature by operating according to 'closed loop cycles'. They have impressive techniques for helping industry use less energy, and can even demonstrate how to use renewables such as wind and solar. They also speak of 'biomimicry', whereby nature's design secrets are put to use in creating products with greatly reduced ecological impacts. These are laudable goals and are to be encouraged and applauded, but to think that the problems of growth will go away because of them is misguided. In the Western world, significant improvements in the efficiency of resource use have not prevented a significant increase in waste and pollution. Even if the supposed hyper-efficiencies do eventually eliminate the problem of material throughput, which I doubt will be possible, we would still be left with the social problems created by the greed and selfishness that we must cultivate in ourselves if the economy is to grow.

The famous phrase 'sustainable development', introduced in 1987 in the Brundtland Report, *Our Common Future*, is meant to address these problems by proposing that more growth is required to generate the wealth to make sustainability a reality, especially in the global South, where it was perceived, quite rightly, that standards of living had to increase. But 'sustainable development' is an oxymoron, as long as 'development' implies increasing the extraction rates of raw materials from wild nature. If so, sustainability and development are contradictory concepts and 'sustainable development' is just economic growth dressed up in the language of deliberate obfuscation, used knowingly or not by those who care nothing for the Earth in order to fool us into thinking that they are taking her concerns seriously.

Steady-State Economics

For it to be truly sustainable, development would be aimed at ensuring that the amount of matter flowing through the global economy would either shrink or be at a *steady state*. The founders of economics, people like Adam Smith and John Stuart Mill, had no problem with steady-state economics. John Stuart Mill wrote, "The best state for human nature is that in which, while no one is poor, no one desires to be richer, nor has any reason to fear being thrust back by the efforts of others to push themselves forward." In modern times a key steady-state economist is Herman Daly, who suggests that a steady-state economy would have four principal characteristics. First, there would be caps set by the best available science for the throughput of every single element and molecule that we wish to incorporate into our production processes, or which we manufacture. These limits will be set at well below the levels that scientists agree can be dealt with by Gaia. This principle has been put to work by American power stations, which began a 'cap and trade' system for trading sulphur emissions up to an agreed total emissions limit. The scheme has reduced sulphur emissions by 30% below the legal ceiling, and cost far less than expected. The same approach could well work for controlling global carbon emissions as long as equity issues are addressed. Second, there would be a market system in place so that the permitted amounts of raw material, or credits for their consumption, can be traded and sold as commodities. The third characteristic is about equity: there must be a limit on how rich any individual business or country can become, and there must be a fair distribution of wealth amongst nation states so that there is no sharp contrast in wealth between very rich and very poor countries. There is an equity principle here which implies that the materially poorer countries of the South would increase their resource use in order to eliminate abject poverty, whilst those of us in the rich North would reduce our own consumption. But overall, the consumption of any given resource would not be allowed to exceed its scientifically determined ceiling. The last requirement for steady-state economics is a stable global population.

Daly's outline for a steady-state economics is very similar to another approach, which avoids the thorny issue of growth because it is designed to be used with the business world, which finds any critique of growth deeply threatening. The approach, known as The Natural Step, was developed in Sweden by Karl-Heinrich Robèrt, a brilliant surgeon who decided that a scientific consensus was needed about how to tackle the ecological crisis. Karl-Heinrich and his eminent team agreed upon four 'system conditions' to which a truly sustainable society would have to adhere. These are:

That nature won't be subject to systematically increasing:
1. Concentrations of substances extracted from the earth's crust;
2. Concentrations of substances produced by society;
3. Degradation by physical means;
   and:
4. That human needs are met worldwide.

The first system condition refers to raw, unprocessed materials that are extracted directly from Gaia, such as metals and fossil fuels, which mustn't be extracted faster than nature can re-absorb and recycle them. The second system condition is about man-made chemicals—DDT, CFCs, artificial fertilisers and so on, which must not be allowed to accumulate in the biosphere faster than they can be broken down and recycled by Gaia into harmless compounds. The third system condition tells us that we cannot continuously degrade wild nature without dire consequences, for it is the 'natural capital' on which our well-being depends. The last system condition recognises that equity is central to sustainability, and that material wealth needs to be fairly distributed within a given society and between the nations of the world. If we were to add an explicit stipulation of steady-state to this list, which as it stands is only implied, and another equally explicit stipulation that the human population should not increase, we would then have a fairly good set of ground rules for living well with Gaia.

An economy organised according to steady-state principles is one that nurtures the soul of place in a way that is consistent with living sustainably within our animate Earth. The growth economy, on the other hand, eradicates the unique qualities of place and literally 're-places' them with the bland, soul-numbing homogeneity demanded by the global economy and the corporations. Global fast food chains replace local restaurants serving local dishes, giant multinational supermarkets replace small shops selling local produce, and local organic farmers using ecologically diverse growing methods are replaced by agribusiness enterprises growing crop monocultures sprayed with toxic chemicals.

Local community is of paramount importance because it is the source of wealth, soul, human warmth and well-being, and so a truly sustainable national economy would have to consist of a network of semi-autonomous local economies operating according to the Gaian rules we have just explored. Profits would mostly be made and circulated locally rather than being siphoned off into the pockets of distant investors, and currencies would be locally created and locally distinctive, decoupled from the official national currency so as to further protect the local economy from adverse outside influences. There are now many experiments operating around the world in which community members offer each other services paid for in a local currency without needing access to the national equivalent.

Food would be grown locally by means of diverse organic production systems that avoid massive fields growing vast acreages of crop monocultures, and that steer clear of cruel forms of animal husbandry. Local people would know exactly where, how and by whom their food was grown, and would be involved in the actual growing to some extent. This kind of practice already exists, and is known as Community Supported Agriculture, or CSA. The most common arrangement is for a local grower to make up 'vegetable boxes' for distribution to community members in exchange for some labour and a guaranteed income. The practice promotes community when local people meet on the farm to contribute their labour, or when they arrive at a vegetable box distribution point, which is often a local household. Helena Norberg-Hodge, director of the International Society for Ecology and Culture (ISEC), and a leading exponent of localisation, believes that strengthening local economies is the single most effective way of resolving our social and ecological crises. "It is a win-win strategy," she says, "for both people and the Earth."

In many parts of the world, locally based, steady-state economies could heal the damage that has resulted from decades of harmful mainstream agricultural practices. In Britain, the countryside has suffered serious degradation since the early 1950s because of a misguided farm subsidy system that has maximised food production at the expense of human and ecological health. The vast majority of the British landscape is now farmed with intensive inputs of chemical fertilisers and pesticides, to the extent that most of Britain is now largely an ecological wasteland with severely diminished biodiversity. Birds once common thirty years ago, such as song thrush, lapwing, house sparrow and starling, are suffering precipitous declines, and many wild plants, insects, mammals and amphibians are faring no better.

In a truly sustainable Britain, these trends could be reversed by means of a seemingly paradoxical policy, namely by encouraging city dwellers to re-create vibrant eco-communities in the countryside. Government subsidies, no longer needed to fund the defunct death-dealing agriculture, would be used to help establish these new eco-communities, whose members would be given financial and technical help with building their own low-impact ecologically sound dwellings out of straw bale, cob or local sustainably harvested timber, which would blend in well with the landscape. For those that wanted it, the new eco-dwellings could be spread out to give space and the opportunity for silent communion with nature. There would be a stipulation that a large proportion of the land around each eco-dwelling should be managed for wildlife or be allowed to regenerate naturally, depending on the characteristics of each site. Every able-bodied community member would be required to contribute a minimum amount of time to working on the community farm, which would be managed by professionals with experience in ecologically diverse organic food production in what would look like a large-scale organic market garden. Many crop species would be grown together in order to benefit from the well-known synergies which on-site diversity brings, such as increased biomass production and pest resistance. Each community would have its own small school, and it would also have its own doctor, bus driver, carpenter and craftspeople. It would care for its own infants, children and elderly people, partly by encouraging rich and friendly interactions between them. It would have its own renewable energy production systems based on solar, wind and biomass. There would be at least one skilled community facilitator/counsellor to help social bonds between community members to mature and grow. Surplus food would be sent to the cities, where local residents would grow additional food in their own communal organic gardens, as has been done with great success in Havana, Cuba. In the cities, many buildings would have been ecologically retrofitted with super-insulation and solar or wind-powered energy generating systems. Corridors of free nature, either as woodland or rough grassland, would weave through the city and out into the wider countryside, where there would be a great network of large, interlinked areas of nature largely free of human interference where every citizen could roam without let or hindrance. As has been the practice amongst humans for countless millennia, adolescents would come of age by means of deep experiences in wild nature guided by qualified adults, in which our young people would encounter the deep mysteries of the cosmos.

I can hear you thinking that all of this probably sounds like a typical hippy-dippy pipe dream—laudable, but totally impractical. Am I really saying that everything will work out just fine if we get ourselves into straw bale houses in the countryside, growing our own vegetables and chickens? Wouldn't it be nice if it were that simple? Alas, it isn't, because we need some drastic action to curb carbon dioxide emissions before the critical thresholds are crossed. In order to achieve this we are going to need a wide range of approaches. Some people will need to pioneer feasible, enjoyable, post-industrial planet-friendly lifestyles. Others will have to work on solving the severe problems of energy, food provision and sea level rise that will come about out as a result of climate change. Others will have to work on the economic arrangements for a steady-state economy, including the reform or dismantling of the major instruments of the war against nature, such as the multinational corporations, the IMF, World Bank, WTO and so on. Ecopsychologists will be needed to engage in the most difficult task of all—how to shift our collective world-view towards the *anima mundi* perspective of experiencing the cosmos as a living being in which we all fully participate. But whilst all of this is going on, we will need to buy ourselves the scarcest commodity of all in this desperate situation: time.

I have major reservations about the strategies suggested by the discipline known as Earth System Engineering, which aims to find high-tech solutions to the problem of climate change. People in this field have proposed all sorts of daring schemes, including putting mirrors out in space to deflect solar energy away from the Earth, stimulating the biological pump by seeding the oceans with iron, or triggering cloud formation by spraying sea salt into the air above the oceans using solar-powered windmills. But there is one option that might work as a band-aid to give us a literal breathing space whilst we work on the full spectrum of tasks to be tackled if we are to live well with Gaia. I am referring to the capture and storage of carbon dioxide as it comes out of the chimneystacks of power stations and factories. Once captured, the gas is pumped into large underground spaces capable of holding it safely in place over geological time scales. The large caverns left over from the extraction of oil and natural gas might work well, and there appear to be enough such places available in the USA alone to hold 100 years of its carbon dioxide emissions. Technically, carbon capture and storage is not a difficult task and is already being done successfully in a handful of pilot projects around the world. STATOIL, the Norwegian state oil company operating in the Sleipner gas field, captures and stores a million tonnes of carbon dioxide every year from natural gas purification in a sandstone aquifer 800 metres underground, thereby reducing Norway's total carbon emissions by 3%. To make a major contribution, the technology will have to be linked to coal-fired power stations, a step that planners seem reluctant to take—the vast numbers of new power stations in the pipeline for India and China (713 in all), have no built-in capacity for carbon capture and storage. There is, however, one coal-fired power station being planned to open in 2015 that will show what this technology can do—the Future Gen plant in the USA, which aims to be a zero-emission facility.

There is of course a great danger that this technology will be used to stimulate even more economic growth by removing all impediments to the use of fossil fuels. If so, the strategy would have failed and the band-aid would have turned into a noose, for, as we have seen, material growth will eventually destroy the biosphere as we know it, and with it our equable climate. There is another related danger of the unwise deployment of this option: our use of it to control global temperature. We could burn some fossil fuels to warm things up if it looked like Gaia was moving towards a new ice age; or we could cool things down if things seemed to be moving in the opposite direction, by capturing and storing the requisite amount of carbon. Either way, we would have failed to live well with Gaia, for humans are not meant to be the masters or stewards of the planet or its bio-geochemical cycles—those are jobs best left to the vastly more climatically competent other-than-human world which surrounds us.

My preferred pathway involves a move towards a global steady-state economy by means of worldwide legal enforcement of energy efficiency, massive investment in renewable energy sources, and a cap and trade scheme for many of the elements and molecules we use, like the one that the European Union and the Kyoto agreement are currently experimenting with for carbon (even though they are setting woefully inadequate limits on emissions of greenhouse gases). We will also need to drastically reduce and in many cases eliminate the insane and energy-hungry practice of transporting produce around the world in favour of local production wherever possible. For these measures to work, the vast majority of citizens around the world will have to understand the fact that Gaia is now beginning to impose the penalties of climate change for the massive transgressions of our industrial lifestyles. Substantial measures and real 'joined-up government' will be needed to make these policies work. So far, this type of governance has been sadly lacking. A recent report by the UK House of Lords Science and Technology Committee showed that current energy-efficiency measures in the UK are ineffective because the money they save is being spent to buy more goods or to expand business activities; because government policies are not requiring strict energy-efficiency standards for products ranging from buildings to electrical goods, and because insufficient funding is going into relevant research.

Deep Ecology

This brings home the point that the real change has to be an inner one, for, as we have just seen, even the most brilliant technological solutions could lead to disaster if they are not used by wise human beings. Perhaps the most fundamental maxim of Gaian wisdom is that we humans are fully accountable to our planet. There is simply no getting around this, for humans are just as subject as any other species to the basic law of Gaia, which, as you will remember, can be drawn like this (Figure 43):

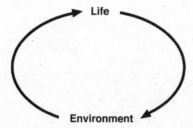

Figure 43: The relationship between life and environment. Note that the arrows in this and the following figure denote relationships, not direct couplings.

When translated into the human realm, the diagram transforms into Figure 44.

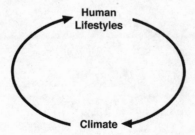

Figure 44. The relationship between human lifestyles and climate.

It's quite a simple piece of wisdom which is worth repeating: *We are all accountable to Gaia*. Any human lifestyle that destabilises a key aspect of the 'environment', such as climate, will be limited by feedbacks curtailing that lifestyle. The beauty of this simple scheme is that it is scientifically undeniable, and that people holding all sorts of philosophical perspectives can accept it. A die-hard mechanist can take it on board just as much as a radical animist, as a guide for action.

But how do we develop ecological wisdom? In my own interpretation of the deep ecology approach pioneered by Arne Naess, each person must work out their own *ecosophy* (from *oikos*: household and *sophia*: wisdom) based on their own deep experience, deep questioning and deep commitment. My friend the Norwegian philosopher and educator Per Ingvar Haukeland has developed a very useful modification of some of Arne Naess's work, to describe how individual ecosophies relate to each other. The basic idea is to represent the situation by means of an *Ecosophical Tree*:

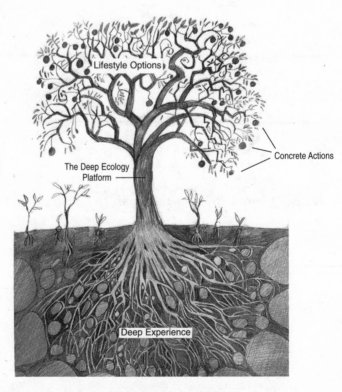

Figure 45: The Ecosophical Tree.

The tree's roots snake down into the rich soil of deep experience, absorbing the nutrition of profound inspiration. Each root tip represents an individual person's deep experience of wide identification with Gaia and with the whole of creation, nourished by his or her own unique part of the soil. These deep experiences need not be fully consistent with each other—a Buddhist would not evoke God as part of their deep experience, but a Christian would, although both would agree about the importance of compassion in our relationships with the whole of life. Naess stresses the importance of *radical pluralism* at this level, for we need to be tolerant of other people's deep experiences, no matter how different they might be from our own.

Commonality arises at the trunk of the tree, into which all the roots flow. Here we encounter the deep ecology platform, a set of eight points formulated by Arne Naess and the American philosopher George Sessions, with which most people inclined towards deep ecology would generally agree.

The Deep Ecology Platform

1. All life has value in itself, independent of its usefulness to humans.

2. Richness and diversity contribute to life's well-being and have value in themselves.

3. Humans have no right to reduce this richness and diversity except to satisfy vital needs in a responsible way.

4. The impact of humans in the world is excessive and rapidly getting worse.

5. Human lifestyles and population are key elements of this impact.

6. The diversity of life, including cultures, can flourish only with reduced human impact.

7. Basic ideological, political, economic and technological structures must therefore change.

8. Those who accept the foregoing points have an obligation to participate in implementing the necessary changes and to do so peacefully and democratically.

(This version of the Deep Ecology Platform was created by participants at Schumacher College in May 1995.)

Notice that the platform begins with a statement from the realm of deep experience—we cannot rationally justify the first point, but we know what it means if it is consistent with our deep sense of belonging to an animate cosmos. The last point is about action, which is what makes deep ecology a movement as much as a philosophy.

Branches begin to spread out in all directions as we move beyond the trunk of the ecosophical tree, representing the options that each person has for making changes in their lives that are consistent with their deep experience of belonging to our animate Earth. Once again, radical pluralism is important here. Different people will make their own unique choices that may well differ very markedly from each other: one person may think about engaging in direct action, whilst another may be considering leaving their high-paying job in the city and joining an NGO. Others will continue to work within the mainstream, striving to change it from the inside. These choices must be respected and supported by other people inclined towards deep ecology.

Living in harmony with the animate Earth involves translating our deep experience of her sentient presence into concrete, everyday actions. Perhaps one of the most important options we can consider, especially if we are living an affluent Western lifestyle, is to find ways of cutting back our personal emissions of greenhouse gases. The IPCC recommends that humanity as a whole needs to cut its emissions by 60% by 2050 relative to 1990 levels—a seemingly impossible task given our immense and increasing hunger for energy and raw materials. But according to Dave Reay from the University of Edinburgh, this level of reduction is possible without too much effort. His 'ten steps to saving the planet' empower us with the idea that seemingly small actions at the personal level can, writ large, have massive consequences. Here are Reay's ten steps:

1. Turn down the heating or turn up your air conditioning by 1^{0}C; this can reduce your carbon emissions by up to 2 tonnes of carbon per annum.

2. Use your car as little as possible—use a bike or public transport whenever you can; potential savings are up to 12 tonnes of carbon per year.

3. Compost your organic waste, thereby denying it to methane-generating bacteria in landfill sites. You could save up to 1 tonne of carbon per year this way.

4. Avoid flying, especially on short-haul flights to destinations reachable by train. If you must fly, you can offset your emissions to some extent by supporting organisations such as Climate Care. Potential savings: up to 3 tonnes of carbon per year.

5. Drive just below speed limits, avoid short journeys, car share and service your vehicle. Think of changing to a diesel, or to a hybrid vehicle. Savings could be as high as 12 tonnes of carbon per annum.

6. Going for energy-saving light bulbs and energy-efficient appliances, and turning them off when you aren't using them could save you up to 1.6 tonnes of carbon per year.

7. Eat locally produced food, thereby eliminating your support for the vastly polluting global transportation network that emits about 80kg of greenhouse gases even for the average shopping basket of *organic* food. This way, you'll save up to 3.6 tonnes of carbon per year.

8. Reduce, reuse, recycle. Before you buy something, ask yourself whether it really is a vital need. Use things many times—such as plastic bags, and, if you can't do the first two, then recycle as much as you can, thereby saving energy for the production of new items. Annual saving: 1 tonne of carbon.

9. Take your deep ecology perspective to your place of work; turn off lights and standbys, print on both sides of the paper, and convince the powers that be to recycle as much as possible.

10. Go for a natural burial and avoid burying concrete, metal and other materials with you, thereby saving up to one tonne of carbon.

I would add an 11th step, which is:

> 11. Spread the word about the urgent need for a steady-state economy before the growth economy makes the previous ten steps irrelevant.

Eventually, after careful consideration, each person will make his or her choices and act concretely in the world. When this happens a fruit grows, matures and finally falls to the ground as soon as the action is carried out, fertilising the soil of deep experience with nutrients that everyone can draw on. Every person's actions will be different, but if they are truly consistent with the realm of deep experience, each action will benefit the whole of Gaia, including other human beings. Each person's particular journey from roots to trunk, and then to branches and fruit represents their own ecosophical path into right action in the world.

The most satisfying actions in this time of ecological and social crisis will be those that are inspired by a sense of service to Gaia that comes from a deep feeling of belonging to the animate community of organisms, air, rocks, and water that constitutes the very fabric of our living planet. We have seen how the science of Gaia is consistent with a deeply intuitive understanding of the Earth as an evolving, ever-changing animate being, and have explored how this new perspective opens up fruitful ways of making peace with the Earth that greatly enrich our rationality and our science. To act well, we need to experience the Earth not as 'nature' out there, nor as an 'environment' that is distinct from us, but as a mysterious extension of our very own sensing bodies that nourishes us with an astonishing variety of intellectual and aesthetic experiences—with the roar of the sea and with the wonderful sight of the night moon reflected in a calm lake. Right action requires us to *live into* the body of the Earth, so that we feel just as comfortable with the air, water, rocks and living beings that are the life of that wider body as we do in our human-made environments. If we could only do this, our focus would shift from the endless fascination with human affairs to a wider, more fulfilling perception of the animate Earth in which these affairs take place. We would then encounter a broader, Earth-centred view in which every breath we take and every decision we make is a pledge of service and allegiance to the greater personhood of our planet. A contribution to this task is to discover new ways of speaking of our scientific insights about the Earth that allow their animate dimensions to emerge, as we have tried to do in this book. With our reason satisfied, our intuition, sensing and feeling are free to forge a connection so deep that we no longer need to think of it. Only when our four ways of knowing are fully engaged in this way can right action emerge—and it is only the summed effect of billions of right actions by people across the planet that may eventually lead us into a genuinely fruitful relationship with Gaia, our animate Earth.

Bibliography

Chapter 1: *Anima Mundi*

David Abram, *The Spell of the Sensuous*, Vintage Books, 1997.

Thomas Berry, *The Great Work*, Crown Publications, 2000.

Henri Bortoft, *The Wholeness of Nature*, Floris Books, 2004.

Fritjof Capra, *The Web of Life*, Flamingo (Harper Collins), 1997.

Margaret Colquhoun, *New Eyes for Plants*, Hawthorn Press, 1996.

A. K. Dewdney, *Hungry Hollow: The Story of a Natural Place*, Springer-Verlag, 1998.

Brian Goodwin, *How the Leopard Changed its Spots*, Weidenfield and Nicholson, 1994.

Stephan Harding, 'What is Deep Ecology', *Resurgence*, No. 185, 1997.

Graham Harvey, *Animism: Respecting the Living World*, Columbia University Press, 2005.

James Hillman, *Revisioning Psychology*, Harper Collins, 1992.

Carl Gustav Jung, *On the Nature of the Psyche*, Routledge, 2001.

Lee R. Kump, James F. Kasting and Robert G. Crane, *The Earth System*, Pearson Prentice Hall, 2nd edition, 2004.

Robert Lawlor, *Voices of the First Day: Awakening in the Aboriginal Dreamtime*, Bear and Company, 1991.

Paul Lowman, *Exploring Space, Exploring Earth*, Cambridge University Press, 2002.

Mary Midgley, *Does the Earth concern us?*, Gaia Circular, Winter/Spring 2002.

Mary Midgley, *Gaia: The Next Big Idea*, Demos, 2001.

Jacques Monod, *Chance and Necessity*, Collins, 1972.

E. F. Schumacher, *Small is Beautiful: Study of Economics as if People Mattered*, Vintage, 1993.

Paul Shepard, *Nature and Madness*, University of Georgia Press, 1998.

Richard Tarnas, *The Passion of the Western Mind*, Pimlico, 1996.

Wemelsfelder, Francoise, 'The scientific validity of subjective concepts in models of animal welfare', *Applied Animal Behaviour Science*, 53, pp.75–88, 1997.

Alfred North Whitehead, *Process and Reality*, corrected and edited by D. R. Griffin and D. W. Sherburne, Free Press, 1978.

Donald Worster, *Nature's Economy*, Cambridge University Press, 1977.

Chapter 2: Encountering Gaia

Timothy Lenton, 'Gaia and Natural Selection', *Nature* Vol. 394, pp.439–447, 1998.

Aldo Leopold, *A Sand County Almanac*, Oxford University Press, 1968.

James Lovelock, *Gaia: A New Look at Life on Earth*, Oxford University Press, 2000.

James Lovelock, *Gaia and the Theory of the Living Planet*, Gaia Books 2005.

James Lovelock, *Homage to Gaia*, Oxford University Press, 2000.

James Lovelock, *The Ages of Gaia*, Oxford University Press, 2000.

James Lovelock, *The Revenge of Gaia*, Allen Lane, 2006.

Maurice Merleau-Ponty, *Phenomenology of Perception*, translated by Colin Smith, Routledge & Kegan Paul, 1962.

Arne Naess, *Ecology, Community and Lifestyle*, translated by David Rothenberg, Cambridge University Press, 1990.

Richard Nelson, *Make Prayers to the Raven: A Koyukon View of the Northern Forest*, University of Chicago Press, 1983.

George Sessions (editor), *Deep Ecology for the 21st Century*, Shambhala, 1995.

Charlene Spretnak, *Lost Godesses of Early Greece*, Beacon Press, 1992.

Laurens van der Post, *The Heart of the Hunter*, Vintage, 2002.

Chapter 3: From Gaia Hypothesis to Gaia Theory

K. C. Condie and R. E. Sloan, *Origin and Evolution of Earth: Principles of Historical Geology*, Prentice-Hall, 1998.

Richard Dawkins, *The Extended Phenotype*, Oxford University Press, 1999.

Jacques Grinevald, 'Sketch for a History of the Idea of the Biosphere' in *Gaia in Action*, edited by Peter Bunyard, Floris Books, 1996.

Stephan Harding, 'Food Web Complexity Enhances Ecological and Climatic Stability in a Gaian Ecosystem Model' in *Scientists Debate Gaia*, MIT Press, 2004.

Richard B. Primack, *Primer of Conservation Biology*, Sinauer Associates Incorporated, 2004

'An Introduction to Systems Thinking', Stella Research Software, High Perfomance Systems Inc, 1997.

Stephen Schneider and Randi Londer, *The Coevolution of Climate and Life*, Sierra Club Books, 1984.

Vladimir I. Vernadsky, *The Biosphere*, translated by D. B. Langmuir, Springer Verlag, 1998.

Chapter 4: Life and the Elements
Philip Ball, *The Ingredients*, Oxford University Press, 2002.
Christian De Quincey, *Radical Nature: Rediscovering the Soul of Matter*,
    Invisible Cities Press, 2002.
John Emsley, *Nature's Building Blocks*, Oxford University Press, 2001.
Timothy M. Lenton and Andrew Watson, 'Redfield Revisited 2:
    What regulates the oxygen content of the atmosphere?', *Global
    Biogeochemical Cycles* Vol 14 No.1, pp.249–268, 2000.
Primo Levi, *The Periodic Table*, Penguin Books, 2000.
Freya Mathews, *For Love of Matter: A Contemporary Panpsychism*,
    SUNY Press, 2003.
David Skrbina, *Panpsychism in the West*, MIT Press, 2005.
Lee Smolin, *The Life of the Cosmos*, Phoenix, 1998.
Brian Swimme, *The Hidden Heart of the Cosmos*, Orbis Books, 1999.
Peter D. Ward and Donald Brownlee, *Rare Earth*, Copernicus, 2000.

Chapter 5: Carbon Journeys
Richard B. Alley, *The Two Mile Time Machine*, Princeton University
    Press, 2000.
R. A. Berner, 'Geocarb II: A revised model of atmospheric CO_2 over
    Phanerozoic time', *American Journal of Science* Vol. 294, pp.56–91,
    1994.
David Schwartzman, *Life, Temperature, and the Earth*, Columbia
    University Press, 1999.
Tyler Volk, *Gaia's Body*, MIT Press, 2003.
Peter Westbroek, *Life as a Geological Force*, W. W. Norton, 1992.

Chapter 6: Life, Clouds and Gaia
Gordon Bonan, *Ecological Climatology*, Cambridge University Press, 2002.
R. Charlson, J. Lovelock, M. Andreae and S. Warren, 'Oceanic
    phytoplankton, atmospheric sulphur, cloud albedo and climate',
    Nature Vol. 326, pp.655–661, 1987.
M. Claussen et al, 'Simulation of an abrupt change in Saharan
    vegetation at the end of the mid-Holocene', *Geophysical Research
    Letters* Vol. 24, pp.2037–2040, 1999.
P. B. deMenocal et al, 'Abrupt onset and termination of the African
    Humid Period: Rapid climate response to gradual insolation forcing',
    Quaternary Science Reviews Vol. 19, pp.347–361, 2000.

W. D. Hamilton and T. M. Lenton, 'Spora and Gaia: How microbes fly with their clouds', *Ethology Ecology and Evolution* Vol. 10, pp.1–16, 1998.

Lynn Hunt, 'Send in the Clouds', *New Scientist*, pp.29-33, 30th May 1998.

Lee Klinger and David J. Erikson, 'Geophysiological coupling of marine and terrestrial ecosystems', *Journal of Geophysical Research* Vol. 102, pp.25359–370, 1997.

Eric Post, Rolf Peterson, Nils Stenseth & Brian McLaren, 'Ecosystem consequences of wolf behavioural response to climate', *Nature* Vol. 401, pp.905–907.

Chapter 7: From Microbes to Cell Giants

Noam Bergman, Timothy Lenton and Andrew Watson, 'A New Biogeochemical Earth System Model for the Phanerozoic' in *Scientists Debate Gaia*, MIT Press, 2004.

Eshel B. Jacob, Israela Becker, Yoash Shapira and Herbert Levine, 'Bacterial linguistic communication and social intelligence' in *Trends in Microbiology*, Vol. 12. No. 8, pp.366–372, 2004.

Lynn Margulis, *Symbiotic Planet*, Phoenix, 1999.

Lynn Margulis and Dorion Sagan, *Microcosmos: Four Billion Years of Microbial Evolution*, University of California Press, 1997.

Humberto Maturana and Francisco Varela, *The Tree of Knowledge*, Shambhala, 1992.

Dorion Sagan, *Biospheres*, McGraw-Hill, 1990.

Vaclav Smil, *The Earth's Biosphere: Evolution, Dynamics, and Change*, MIT Press, 2003.

Chapter 8: Desperate Earth

P. M. Cox, R. A. Betts, C. D. Jones, S. A. Spall and I. J. Totterdell, 'Acceleration of global warming due to carbon-cycle feedbacks in a coupled climate model', *Nature*, 408, pp.184–187, 2000.

W. Steffen et al, *Global Change and the Earth System*, Springer, 2004.

Robert T. Watson (editor), *Climate Change: Contribution of Working Groups I, II, III to the Third Assessment Report of the Intergovernmental Panel on Climate Change*, Cambridge University Press, 2002.

Chapter 9: Gaia and Biodiversity

Gregory Bateson, *Mind and Nature*, Hampton Press, 2002.

Rachel Carson, *Silent Spring*, Pelican Books, 1998.

Hector et al, 'Plant diversity and productivity experiments in European grasslands'. *Science*, 286, pp.1123–1127, 1999.

Michel Loreau (editor), *Biodiversity and Ecosystem Functioning*, Oxford University Press, 2002.

E. O. Wilson, *Biophilia*, Harvard University Press, 1984.

E. O. Wilson, *The Diversity of Life*, Allen Lane, 1992.

E. O. Wilson, *The Future of Life*, Little Brown, 2002.

Chapter 10: In Service to Gaia

Brad Allenby, 1999, 'Earth System Engineering: The Role of Industrial Ecology in an Engineered World', *Journal of Industrial Ecology*, Vol. 2 Issue 3, pp.73–93.

Janine Benyus, *Biomimicry*, Harper Collins, 2002.

Fritjof Capra, *The Hidden Connections: A Science for Sustainable Living*, Flamingo, 2003.

David Cook, *The Natural Step* (Schumacher Briefing No. 11), Green Books, 2004.

Herman Daly, *Steady-State Economics: The Economics of Biophysical Equilibrium and Moral Growth*, W. H. Freeman, 1978.

Richard Douthwaite, *The Growth Illusion*, Green Books, 1999.

Clive Hamilton, *Growth Fetish*, Pluto Press, 2004.

Paul Hawken, Amory Lovins and Hunter Lovins, *Natural Capitalism*, Earthscan, 2000.

Craig Holdrege, *Genetics and the Manipulation of Life: The Forgotten Factor of Context*, Lindisfarne Books, 1996.

Jerry Mander and Edward Goldsmith (editors), *The Case Against The Global Economy: And for Local Self-Reliance*, Earthscan, 2001.

Donella Meadows, Jorgen Randers and Dennis Meadows, *Limits to Growth: The 30-Year Update*, Chelsea Green, 2004.

Helena Norberg-Hodge, Todd Merrifield and Steven Gorelick, *Bringing the Food Economy Home*, Zed Books, 2002.

David Raey, *Climate Change Begins at Home*, Macmillan, 2006.

Ernst Von Weizsacker, Amory Lovins and Hunter Lovins, *Factor Four*, Earthscan, 1998.

Index

www.waterworks.com.au
(Mushroom fountain)